T0320170

Effective Global Carbon Markets

ELGAR STUDIES IN CLIMATE LAW

Series Editor: Jonathan Verschuuren, *Professor of International and European Environmental Law, Tilburg University, the Netherlands*

Climate change and responses to climate change are having an increasing effect on economic activities and peoples' lives. Across the globe, scholars are studying what these impacts mean or should mean for law. As a consequence, climate law is rapidly developing as a new and complex field of law, marked by the interactions of diverse areas such as migration law, human rights law, agricultural law, energy law, trade law, company law, tort law, insurance law, nature conservation law, marine law and so forth. This series publishes high quality scholarly works that make an essential and innovative contribution to the development of legal thinking on climate law and thus help us to understand how we can protect tomorrow's society against the consequences of climate change.

Titles in the series include:

The Concept of Climate Migration
Advocacy and its Prospects
Benoît Mayer

Climate Change and Catastrophe Management in a Changing China
Government, Insurance and Alternatives
Qihao He

Effective Global Carbon Markets
Networked Emissions Trading Using Disruptive Technology
Justin D. Macinante

Effective Global Carbon Markets

Networked Emissions Trading Using Disruptive Technology

Justin D. Macinante

Post Doctoral Researcher, Law School, University of Edinburgh, UK

ELGAR STUDIES IN CLIMATE LAW

Edward Elgar
PUBLISHING

Cheltenham, UK • Northampton, MA, USA

Published by
Edward Elgar Publishing Limited
The Lypiatts
15 Lansdown Road
Cheltenham
Glos GL50 2JA
UK

Edward Elgar Publishing, Inc.
William Pratt House
9 Dewey Court
Northampton
Massachusetts 01060
USA

A catalogue record for this book
is available from the British Library

This book is available electronically in the **Elgar**online
Law subject collection
http://dx.doi.org/10.4337/9781839109485

ISBN 978 1 83910 947 8 (cased)
ISBN 978 1 83910 948 5 (eBook)

Printed and bound by CPI Group (UK) Ltd, Croydon, CR0 4YY

Contents

Figures and tables

FIGURES

TABLES

Acknowledgements

This book is based on research undertaken at the University of Edinburgh and thus, I begin by thanking all those who helped me with that process. First, many thanks to Navraj Singh Ghaleigh and Professor Burkard Schafer, of Edinburgh Law School, for their wise and timely counsel. Thanks also to everyone in the Law School Postgraduate Research office for their excellent support. My thanks also to Professor Michael Mehling of MIT and Dr Matthew Brander of Edinburgh University Business School, for their valuable comments on the work.

The evolutionary process to get to the point of publication of this book began much earlier and has been more involved than just the researching and writing of the work. It involves a long list of actors, all of whom I thank, but some of whose roles I wish to acknowledge here, in particular.

The book elaborates a model for internationally networking carbon markets using distributed ledger technology, aimed at making carbon pricing more effective in enhancing mitigation efforts globally. The networking concept originated in the World Bank's Networked Carbon Markets initiative, to which I was introduced by my long-time friend Martijn Wilder, to whom special thanks is given. Martijn's knowledge of all matters related to climate change is as comprehensive as his network of contacts. Thanks also to Bianca Sylvester and Chandra Shekhar Sinha, for their great support.

I also thank Professor Gerard C. Rowe, Viadrina European University, Frankfurt (Oder) (Germany) for his constant encouragement, thoughtful and provoking comments, and valuable professional advice throughout many years, and hope this will continue into the future.

Mention is made also of my former colleagues in Frankfurt and Zürich at First Climate AG, with whom many interesting experiences in the carbon market were shared over a number of years and, with some, continue to be shared. All valuable grist to the mill.

Of course, the first roles in this evolutionary process for which acknowledgement and thanks are due, are those of my parents, Joseph and Teresa, who made many sacrifices to ensure that my siblings and I had the benefit of a good education and start in life. Finally, the biggest thank you to Helen, my wife, supporter, critic, sounding board, friend and partner in all life's endeavours.

Abbreviations

AAU	Assigned amount unit
AGBM	Ad Hoc Group on the Berlin Mandate
AML	Anti-money laundering
ASIC	Australian Securities and Investments Commission
AUSTRAC	Australian Transactions Reports and Analysis Centre
AWGPA	Ad Hoc Working Group on the Paris Agreement
BIS-CPMI	Bank of International Settlements – Committee on Payments and Market Infrastructures
BoE	Bank of England
CAD	Compilation and accounting database
CDM	Clean Development Mechanism
CDMEB	Clean Development Mechanism Executive Board
CER	Certified Emission Reduction
CfD	Contract for difference
CHF	Swiss Franc
CMA	Conference of Parties to the Convention, as the Meeting of Parties to the Paris Agreement
CMP	Conference of Parties to the Convention, as the Meeting of Parties to the Kyoto Protocol
CO_2	Carbon dioxide
CO_2e	Carbon dioxide equivalent
COP	Conference of Parties
CORSIA	Carbon Offsetting and Reduction Scheme for International Aviation
CP	Commitment period
CPR	Commitment period reserve
CRA	Credit reference agency
CSDR	Central Securities Depositories Regulation

CTF	Counter terrorism financing
DAG	Directed Acyclic Graphs
DLT	Distributed ledger technology
DNA	Designated National Authority
DOE	Designated Operational Entity
EBA	European Banking Authority
EC	European Commission
ECA	European Court of Auditors
ECC	Elliptical Curve Cryptography
ECDSA	Elliptical Curve Digital Signature Algorithm
EEA	European Economic Area
EIT	Economy in transition
ERU	Emission reduction unit
ESG	Environment, sustainability, governance
ESMA	European Securities and Markets Authority
ETS	Emissions Trading Scheme
EU	European Union
EUA	European Union Allowance
EUETS	European Union Emissions Trading Scheme
EUTL	European Union Transaction Log
FATF	Financial Action Task Force
FCA	Financial Conduct Authority
FINMA	Swiss Financial Market Supervisory Authority
FMLC	Financial Markets Law Committee
FSB	Financial Stability Board
G8	Group of eight highly industrialised countries
G20	Group of twenty major economies
G77/China	Group of 77 developing countries with China
GATS	General Agreement on Trade in Services
GDPR	General Data Protection Regulation
GEF	Global Environment Facility
GFIN	Global Financial Innovation Network
GHG	Greenhouse gas

H_2O	water
HASH	Cryptographic hash function
HE	Homomorphic Encryption
HFC	Hydrofluorocarbon
HMRC	Her Majesty's Revenue and Customs
ICAO	International Civil Aviation Organisation
ICAR	International carbon asset reserve
ICO	Initial coin offering
IET	International emissions trading
ILC	International Law Commission
IMF	International Monetary Fund
INDC	Intended nationally determined contributions
IOSCO	International Organisation of Securities Commissions
IPCC	Intergovernmental Panel on Climate Change
ISDA	International Swaps and Derivatives Association
ISP	International settlement platform
IT	Information technology
ITL	International Transaction Log
ITMO	Internationally transferred mitigation outcome
KP	Kyoto Protocol
LULUCF	Land use, land use change, and forestry
MAD	Market Abuse Directive
MiFID	Markets in Financial Instruments Directive
MOP	Meeting of Parties
MRV	Monitoring (or measurement), reporting, verification
MV	Mitigation value
NCM	Networked carbon markets
NDC	Nationally determined contribution
NZ	New Zealand
OECD	Organisation for Economic Cooperation and Development
OTC	Over-the-counter
P2P	Peer-to-peer

PA	Paris Agreement
PAWP	Work Programme under the Paris Agreement
PBOC	People's Bank of China
PKI	Public key infrastructure
PoA	Programme of Activities
ppm	parts per million
PRA	Prudential Regulatory Authority
PRC	People's Republic of China
QELRC	Quantified emission limitation and reduction commitment
QELRO	Quantified emission limitation and reduction obligation
REC	Renewable Energy Certificate
RGGI	Regional Greenhouse Gas Initiative
RMU	Removal unit
SBI	Subsidiary Body for Implementation
SBSTA	Subsidiary Body for Scientific and Technological Advice
SFC	Securities and Futures Commission (Hong Kong)
SFD	Settlement Finality Directive
SGX	Software guard extensions
SHA-256	Secure hash algorithm
SMSG	Securities and Markets Stakeholder Group
SSL	Secure socket layer
T+1, (2, 3)	Trading date plus one day (two, three days)
TGE	Token generating event
TT	trusted technologies
TU	Transaction unit
UK	United Kingdom
UN	United Nations
UNCED	United Nations Conference on Environment and Development
UNEP	United Nations Environment Programme
UNFCCC	United Nations Framework Convention on Climate Change

UNGA	United Nations General Assembly
US	United States
USEPA	United States Environmental Protection Agency
VAT	Value added tax
VCLT	Vienna Convention on the Law of Treaties
WCI	Western Climate Initiative
WEF	World Economic Forum
WMO	World Meteorological Organisation
WTO	World Trade Organisation
zk-SNARK	zero-knowledge Succinct Non-interactive Arguments of Knowledge

PART I

Introductory matters and background

1.   Introduction to effective global carbon markets

'Time is of the essence' is an expression familiar to lawyers as a typical contractual stipulation. However, considered from the perspective of the state of the global environment, it may be more appropriate to see it as a stipulation for the global social contract – and as a simple statement of fact. Time is indeed of the essence for action to reduce human impact on the global climate system. As intergovernmental negotiations to determine rules to give operational effect to the Paris Agreement[1] stretch on now into a new decade, with the likelihood of more equivocation than clarity, more obstruction than focused engagement, the impression created by diverse political dissemblers might be of abundant time to take any necessary steps, that there is, in fact, no urgency. Yet the critical need for timely action emerges not just from what the world's climate scientists tell us ever more frequently, compellingly, and with ever greater certainty. As the impacts of climate change become more frequent, widespread, and increasingly destructive, urgency is patent to any observer.

In these circumstances, this book seeks to contribute in a specific way to timely mitigation action. It proceeds on the basis that carbon pricing and, in particular, carbon emissions trading, is an important component of mitigation policies in general, and one adopted already in many jurisdictions. The proposal detailed here is of a model that can connect those jurisdictions, so that their policies and programmes are more effective and operate more efficiently.

The model proposes networking of carbon markets through application of distributed ledger technology (DLT). The proposal sets out the infrastructure, operational mechanisms, and rules (including indicative transactional contract terms) that this networking might entail. Importantly, it also advances a governance structure for such networked markets within the policy and regulatory setting of the Paris Agreement. The hypothesis on which the proposal relies is that networking can address issues that previously militated against the effectiveness of international emissions trading as an essential tool of climate policy. The trading mechanism proposed, based on application of the particular

---

[1]   Paris Agreement, FCCC/CP/2015/10/Add.1 <http://unfccc.int/resource/docs/2015/cop21/eng/10a01.pdf> accessed 13/03/17.

technology, has the potential to facilitate better outcomes, thereby expanding positive participation of a range of economic actors.

Thus, the overriding aim of this book is to elaborate a specific operational design for an inter-jurisdictional carbon market, for which currently there is no feasible, effective, or practical equivalent. However, I begin with a short story: as part of my research for this work, I contacted the UK Financial Conduct Authority (FCA) to discuss the International Organisation of Securities Commissions (IOSCO), since the FCA chairs the IOSCO Fintech Network and so is relevant to two aspects of the research – financial market governance; and application of the technology. Even though my call concerned potential financial supervisory governance by IOSCO in relation to a networking of markets on distributed ledger technology, the FCA began by announcing they had their 'ESG person'[2] on the call and, in our discussion, emphasized how important ESG was in their IOSCO work. While this was all very good, it was not the financial market governance element, nor even the idea of this technology being applied to market operation that grabbed attention. What struck me was that the mere mention of climate and carbon seemed to have pigeonholed my call in the ESG box.

The point of this story is to underscore two important aspects of this book, one substantive and one methodological: first, a substantive theme is that climate change is not just an environmental concern, but (inter alia) an economic and financial issue and, as such, needs to be treated as part of the mainstream political-economy debate, rather than as a separate side issue.[3] It is not just (or sometimes, not even) an issue for the 'ESG person'. The operation of carbon markets, domestically and networked globally, should be core business for financial regulators and financial supervisory bodies.

From the methodological perspective, the experience of this call highlights the difficulty of getting real-world engagement on a conceptual model being proposed to operate using a new, innovative technology, for which there are few currently operational applications. Thus, research for this work has been primarily desktop, as opposed to empirical. Moreover, the conceptual and technological basis requires elaboration and explanation, necessitating a certain level of descriptive material to facilitate analysis of the proposal. For instance, background on what is networking of carbon markets, what it entails, how could it come about, and its viability as an alternative to linking in connecting carbon markets, needs to be presented before examining questions

---

[2]   Their staff member responsible for environmental, social, and governance matters.

[3]   'It needs to be released from a compartmentalized framing': Cinnamon Carlarne, 'Delinking International Environmental Law & Climate Change' (2014) 4 *Michigan Journal of Environmental & Administrative Law* 48.

such as whether a networked carbon market could operate within the ambit of the Paris Agreement and, if so, how, or what existing institutional and regulatory frameworks provide models for a future networked carbon market.

Similarly, background information on what DLT is and why it will facilitate networking of carbon markets needs to be elaborated before considering questions such as what the legal issues are to which DLT architecture, as applied to a network of carbon markets, gives rise, and how might those issues be researched, analyzed, and addressed. Before elaborating, however, it is important to first explain briefly where this book sits in the universe of climate change research and writing.

CONTEXT OF RESEARCH

Societal responses to climate change are often categorized as either mitigation, which involves combatting the causes of climate change, or as adaptation, which denotes adapting to the impacts of climate change. This book is looking only at mitigation, measures for which can include regulation, voluntary reductions, subsidies, and education. In turn, regulation can include command and control measures, which prohibit or limit activities, and pricing mechanisms, which aim to influence behaviour via price signals. Pricing mechanisms will need to work in conjunction with a regulatory structure, and taxation and emissions trading are two such pricing mechanisms that sit within this overall framework.

Greenhouse gas (GHG) emissions are not geographically fixed, but pervade the atmosphere, while sources of those emissions are ubiquitous, so the logic of international emissions trading is that mitigation action can occur where it is more cost effective to do so. Thus, if country A and country B both require emitters to reduce emissions below a threshold, or otherwise offset any excesses above their respective levels, then an emitter in country A that can reduce more cheaply can sell any over-achieved reductions to an emitter in country B for which it may be more expensive to reduce than to offset. In this way, both achieve their respective prescribed emission reductions, but with greater overall economic efficiency. This book is about how opportunities for emissions trading to make a difference, in effecting mitigation outcomes, might be maximized.

PURPOSE, OBJECTIVE, THEORY, AND THEMES

1.   Purpose and Objective

The principal purpose of this book is to propose the design and operation of a market (or markets), as a mechanism for implementing the policy of mitigating GHG emissions to help achieve the objective of global climate change policy, being stabilization of GHG concentrations in the atmosphere at a level that would prevent dangerous anthropogenic interference with the climate system.[4] The objective is to arrive at a suitable design for regulatory and institutional frameworks for trading emissions in the context of the Paris Agreement and wider governing structures, treating emissions trading as a financial market and using DLT architecture to connect different markets in a network. In broad terms, the book proceeds to this objective by:

- examining current arrangements for emissions trading, taking account of emissions trading's dual functions as both GHG mitigation policy measure and trading market, from three perspectives: first, a high-level macro-perspective (compartmentalisation of international emissions trading under the Kyoto Protocol[5] in the climate regime); second, a more granular perspective (looking at the nature of what is traded); then third, considering the current state of the carbon market and how it might develop going forward in the context of the Paris Agreement;
- proposing a model for networking carbon markets on a distributed ledger (DL) architecture that addresses both the identified shortcomings and future requirements if the carbon market is to be effective as an instrument of GHG mitigation policy; and
- proposing a framework for analysis of the regulatory and institutional frameworks for such a market, then applying that framework to analyze the proposed governance structure.

2. Theory and Themes

This book envisages a future where there may be trading between the heterogeneous emissions trading schemes (ETSs) and other pricing mechanisms being implemented by jurisdictions, under Article 6, Paris Agreement. It proceeds on the basis that such trading across schemes and jurisdictions can foster larger, deeper, and more liquid markets, less susceptible to manipulation and that more effectively price carbon emissions. The networking approach proposed in this book to achieve such inter-jurisdictional trading is not only a mechanism for implementing the policy of mitigating GHG emissions, but also can be seen as an opportunity for addressing the need for better integration

---

[4]   Article 2 (Objective), United Nations Framework Convention on Climate Change, 9 May 1992, 1771 UNTS 107 (1994); also see recitals, Paris Agreement (fn.1).
[5]   Kyoto Protocol to United Nations Framework Convention on Climate Change, 11 December 1997, 2303UNTS162 (2005).

and mainstreaming of international climate change policy, law, and practice in economic planning globally.

The networked market – if well-designed – should operate as a global financial market, with as little intervention as possible to facilitate its efficient operation and effectiveness, but should do so within an equally well-designed boundary framework of climate change rules that give effect to the intention of the parties to the Paris Agreement. If this can be achieved, it is postulated opportunities for emissions trading markets to make an impact on GHG emissions in an expeditious manner will be greatly enhanced.

A number of themes course through the book. The first of these is that cocooning the carbon market in the climate regime has only perpetuated and, unless changed, will continue to perpetuate a perception that climate change is an issue to be addressed, regulated, and managed separately from and outside the mainstream of national and international economic and financial activity. Second, there is the theme of heterogeneity as opposed to homogeneity – the need for acceptance and recognition of diversity of national approaches according to circumstances and capacities and of the units traded in different schemes: consequently, the need for placing a value on differences, rather than attempting to coerce homogeneity across jurisdictions and schemes.

Third, there is a need for any policy or measure involving the carbon market to engage the private (financial) sector at scale to achieve the best possible outcome. There is a role for both public policymakers and the private sector, it being critical to create an appropriate environment in which each seeks to optimize outcomes.[6] Fourth, the theme that, as such, there is a need for appropriate design of the regulatory and institutional frameworks to promote such outcomes, this not being something that can be left just to multilateral intergovernmental negotiations to draw out to a sub-optimal compromise outcome.

APPROACH AND STRUCTURE

The book begins with a short background, including scientific origins and the intergovernmental responses leading to the Convention and Kyoto Protocol. The policies and measures put in place are briefly considered, as is implementation, particularly through international emissions trading (Chapter 2). Reasons are outlined as to why international emissions trading (IET) can be seen to have failed to achieve its promise. IET – the carbon market that devel-

---

[6]   See, for instance: European Bank for Reconstruction and Development (EBRD), Operationalising Article 6 of the Paris Agreement: Perspectives of developers and investors on scaling-up private sector investment, May 2017 <www.ebrd.com> accessed 21/09/17.

oped under the Kyoto Protocol – is both an environmental policy measure for mitigating greenhouse gas (GHG) emissions, and a financial market, operating at an international level.

Part II of the book picks up that duality of functions performed by IET and on themes including its role as financial market, cooperation between regulators, and the nature of allowances. It examines the carbon market from three perspectives: first, considering the dual functions from a broad, macro-perspective (Chapter 3), second, at a granular level, focusing on the nature of what is traded (Chapter 4) – both these being in terms of how the carbon market has evolved – then third, the carbon market is examined as it stands now and the direction it might take going forward (Chapter 5).

Part III sets out the market proposed to facilitate inter-jurisdictional trading under the Paris Agreement. It does so in two chapters, addressing the concept and theory of the proposal, analyzing it in terms of its component parts, net-worked carbon markets (NCM) on a DLT platform (Chapter 6). It then sets out the proposal elements in practical detail – explaining how it is envisaged the market would operate in practice (Chapter 7).

Part IV establishes the governance structure for the proposed market and a framework for analysis, and then applies that analytical framework. It does so over two chapters, the first dissects the governance structure, then examines it in relation to each of the three areas of law – climate change; financial market regulation; and regulation of the technology and its applications – according to the particular requirements of each (Chapter 8). The second focuses more specifically on the regulatory frameworks, examining in particular the point of their intersection in what is traded, the 'mitigation outcome/financial instru-ment/token', before continuing the analysis by examining legal issues arising from each area relevant to the governance structure (Chapter 9).

Part V (Chapter 10) sets out conclusions. The three areas of climate change law, financial markets regulation, and regulation of distributed ledger technology applications that coalesce in the model and are addressed by the governance structure are constantly evolving and developing. Accordingly, it is acknowledged that inevitably pronouncements, events, and developments may have occurred between the time of completing writing and publication.

INTRODUCTION TO CONCEPTS AND DEFINITIONS

Concepts, technologies, and terms are described or defined as appropriate where used throughout the text. Nevertheless, there are some of fundamental importance that need to be elaborated at the outset, as a proper understanding of their meaning and use is integral from the start. Thus, 'carbon market' and 'distributed ledger technology' are introduced, and uses of some other terms explained.

The carbon market, at present, is profiled in the first section of Chapter 5, but the expression 'carbon market' is used in places in the book as a broad collective description of the various different forms of carbon pricing. It is acknowledged this usage is somewhat loose, since there are forms of carbon pricing, such as carbon taxes, or renewable energy credits (RECs), which would not automatically be associated with a market. The thinking behind this usage is that, if an acceptable methodology could be devised to standardize the mitigation values of all the different types of units, whether allowances, carbon credits, tax credits, RECs, and so on, then (for the moment leaving to one side how this might work in practice) they might all, in future, be capable of being part of a trading mechanism. So, in this broadest, idealized sense, the carbon market might encompass ETSs, carbon credit generating projects, carbon taxes, both the voluntary emission reduction market and compliance market, industry-based schemes (such as that for international civil aviation), and even internal carbon pricing undertaken by corporations.

Notwithstanding this occasionally generalized usage, in the sense used in Chapters 2 and 3, carbon market is referring principally to IET under the Kyoto Protocol, including trading in assigned amount units (AAUs) and project-generated credits, mainly certified emission reductions (CERs); and the European Union ETS (EUETS). Chapter 4 examines the debate about the nature of what is traded and in this sense focuses on allowances in ETSs, as distinct from, for example, credits generated by projects.[7] Chapter 5, second and third sections, focus on ETSs in relation to connecting across jurisdictions. The proposed model market (Part III) and analysis of the governance structure (Part IV) deal only in terms of networking ETSs. This is to reduce complexity and for clarity in explaining the model. Finally, in relation to this expression, 'carbon' market is used to delineate the market described as relating solely to units of GHG mitigation and (at least for present purposes) these would be limited to the GHGs listed in Annex A of the Kyoto Protocol. This does not mean that the market model proposed could not also eventually accommodate units relating to other GHGs or even units of co-benefits,[8] especially once there

---

[7]   An ETS allowance – issued either for free or auctioned – permits emission of a unit of GHG and entities with obligations in the ETS will be required to acquit their emissions by surrendering an equivalent number of allowances, whereas a credit is generated by effecting mitigation, removal, or avoidance of the equivalent amount of emissions that would otherwise occur. Allowances are capped and theoretically reducing over time, whereas the number of credits generated depends on the project.

[8]   Co-benefits include related benefits, such as sustainable development benefits, that may flow from mitigation actions, such as access to clean water, access to electricity, employment opportunities, and so on.

is methodology to provide for credits (such as pursuant to Article 6.4 Paris Agreement) to be traded.

'Distributed ledger technology (DLT)', or 'blockchain', as it is sometimes known (see below), is essentially a bringing together of developments in several digital research fields, such as cryptography and decentralized computer networks. It covers a wide range of potential functionality, so it is useful to identify key features that define what is usually thought of as a DLT system. First, it will be decentralized, meaning that it is made up of multiple computers (also referred to as nodes)[9] and it will be a distributed infrastructure, meaning that instead of there being a central ledger holder through which transactions must pass and where information is recorded, the ledger is held by multiple, or even all nodes in the network. Second, the participants will use encryption to interact with transactions in the system, so there will not need to be a trusted central counterparty. Third, there will be a mechanism by which the nodes reach consensus on the valid entries to add to the ledger. Finally, the ledger entries will be accumulative, so that once they are entered they cannot be altered or removed unless all participants agree, thus the ledger is frequently described as immutable.[10]

There are also elements that are configurable to suit the desired design. These include who has permission to view the ledger or parts thereof and who has permission to alter (add transactions to) the ledger; the transactional terms and conditions embedded in computer code (referred to as 'smart contracts'); and also the arrangements for settlement, exchanges, and payment systems. Configuration of all the elements can result in very different outcomes, all of which could be a DLT system. Blockchain is, in fact, just one type of DLT implementation in which transactions are collected in blocks of information and, once there is consensus, the block is cryptographically linked to the preceding block in the chain of blocks that makes up the ledger.

The term 'fragmentation' is used in two contexts herein. First, in Chapter 3 fragmentation refers to the phenomenon of fragmentation in international law and especially international environmental law that stems from the specialized rules, institutions, and practices that have grown up over time. Then, in Chapter 5, the nature of the global carbon market is described as being fragmented, referring to the diverse and heterogeneous carbon pricing mechanisms (including ETSs) that are being implemented in jurisdictions around the world.

A further clarification relates to the meaning of references made to international emissions trading/the carbon market 'operating as a global financial

---

[9]   Obviously, the computers will all need to be on a network, e.g., connected with each other via the internet.

[10]   It is not quite accurate to call it immutable, as explained in Chapter 6.

market'. Broadly, this is intended to suggest that the market be regulated along the same lines as financial markets, for instance, by the assets traded being defined as financial instruments so as to invoke investor protection, anti-fraud, anti-market manipulation, counter-terrorism financing, anti-money laundering, and systemic risk-management provisions. As an element of these measures, that trading also should take place on a regulated trading platform, or platforms.

The final conceptual/definitional point to introduce here relates to governance and regulation. As explained in Chapter 8, governance (and thus, the expression 'governance structure') is used in a very broad sense. It includes regulatory and institutional frameworks, but also more informal rules and actors, for instance, market discipline as a corrective influence on inaccurate market information (i.e., mitigation value assessments) is included as one of the tiers of the governance structure. Regulation is also used in a generalized sense, referring to laws generally that constrain behaviour of market participants, rather than in the stricter sense of meaning secondary or subordinate legislation that implements primary legislation, unless the context indicates otherwise.

SOME QUALIFICATIONS AND CLARIFICATIONS

This book does not analyze the effectiveness of emissions trading as a mitigation policy measure in comparison to other mitigation measures, nor as a climate policy in relation to climate policies and measures more generally. Nor is it argued that emissions trading is the only mitigation policy measure that should be pursued, or even, the primary mitigation measure.

Climate change is of its nature multi-disciplinary, but just as this does not require this work to engage in atmospheric physics and chemistry proofs to establish the existence of the phenomenon, nor does it require revisiting the economic theory of emissions trading from first principles. Thus, a number of assumptions are relied upon, including for instance, in relation to the economic theory, that regulation with emissions trading can bring about emission reductions more cost effectively and economically efficiently than regulation without emissions trading. Also, that smaller, separate markets operate more efficiently and derive other benefits by connecting with each other to form larger markets. Further, in relation to the technology, it is assumed also that the DLT application proposed herein is technically viable.

Although it may touch on a number of these subjects, the book is not about the global financial regime or global environmental governance, per se, but rather only in so far as emissions trading as a climate change mitigation policy measure touches on them. Nor (as noted above) is it about the economic theory of emissions trading, or why connecting markets is more efficient. It is not about trade (in the sense of the World Trade Organization) and environment, but it is about using a trading mechanism as a policy measure to better achieve the objectives of climate policy. The multi-disciplinary nature of climate change as a global problem, and of the measures to address it, means that it is not possible to avoid touching on political, economic, scientific, technological, social, and developmental aspects, as well as legal. Yet this book is about the law and thus focuses on regulatory and institutional frameworks that may be appropriate to facilitate maximization of the opportunities for emissions trading to make a difference.

2. Background: the problem of and response to climate change

GROWTH IN ANTHROPOGENIC EMISSIONS CAUSING DANGEROUS CLIMATE CHANGE

The question of whether heat absorption by atmospheric gases leads to higher earth surface temperatures was a topic of scientific research from the early nineteenth century. While Svante Arrhenius is sometimes credited with first identifying this so-called greenhouse effect, Arrhenius himself noted that in the 1820s Joseph Fourier had maintained that the atmosphere acts like the glass of a hothouse[1] (thus, suggesting the source of the analogy). In his paper of 1896, Arrhenius also acknowledged the work of John Tyndall in this respect (identifying water vapour, H_2O, as being of the greatest influence, but also the role of carbonic acid gas – now known as carbon dioxide, CO_2) and the research of many others over the course of that century.[2]

Like Tyndall, Arrhenius was interested in what had caused the ice ages, but his focus was on CO_2. His calculations led him to an estimate of the probable effect of a variation in atmospheric CO_2 on the temperature of the earth, that 'if the quantity of carbonic acid increases in geometric progression, the augmentation of temperature will increase nearly in arithmetic progression'.[3] Doubling carbonic acid would increase the temperature by 5–6°C.[4] Thus, he posited that during the ice ages the concentration must have been about half its value at the time of his research. As to whether such was probable, Arrhenius cited

---

[1] Svante A. Arrhenius, 'On the Influence of Carbonic Acid in the Air upon the Temperature of the Ground' (1896) Series 5, Volume 41 *The London, Edinburgh, and Dublin Philosophical Magazine and Journal of Science*, 237–76 <http://www.rsc.org/images/Arrhenius1896_tcm18-173546.pdf> accessed 30/01/19.

[2] Ibid, 238–9.

[3] Ibid, 267.

[4] Ibid, Table VII, 266. This estimate was revised to around 4°C in a later paper: S. A. Arrhenius, 'The Probable Cause of Climate Fluctuations' (1906) *Meddelanden från K. Vetenskapsakademiens Nobelinstitut* Band 1 No 2, (Friends of Science Translation) <https://www.friendsofscience.org/assets/documents/Arrhenius%201906,%20final.pdf> accessed 30/01/19.

Arvid Högbom, whose work identified processes by which carbonic acid was supplied to the atmosphere and was consumed from it. Although he focused on other geological and geochemical processes, interestingly, Högbom mentioned both the role of 'modern industry' in supplying carbonic acid (although he considered this all consumed in the formation of limestone or other mineral carbonates) and the moderating role of the oceans through absorption and evaporation.[5] It would be another half century before these two elements of the cycle were more thoroughly and accurately examined.

In the 1950s, the Cold War provided the impetus and atmospheric nuclear tests provided the opportunity to redress analytical and data shortcomings that had abetted scepticism of Arrhenius's theory and calculations: Cold War research into atmospheric absorption of infrared radiation confirmed the role of CO_2, as distinct from water vapour, in the upper atmosphere;[6] while analysis of tree rings, for the ratio of carbon-14 (the radioactive isotope present due to nuclear testing) to other carbon isotopes, showed increasing percentages of non-radioactive isotopes, indicating the carbon added in newer rings came from fossil carbon sources.[7]

Other research in the late 1950s identified more clearly the role of ocean absorption and evaporation in the carbon cycle; the steady rise of atmospheric CO_2 (evidenced by the Keeling curve); and the consistent rise in baseline temperatures, while by the late 1950s/early 1960s scientists were beginning to sound warnings about the rate of industrial production emissions.[8] Further evidence came from Antarctic ice core measurements going back 400,000 years – through four complete glacial cycles – showing a CO_2 range of 180 parts per million (ppm) to 280ppm over all that period: at the time of this analysis, atmospheric CO_2 concentrations had reached 350ppm.[9]

Returning to Arrhenius and Tyndall, fluctuations in atmospheric CO_2, it transpires, were not the cause of the ice ages, but rather part of a feedback loop leading to the glacial cycles, the trigger being tiny shifts in the Earth's orbit of the Sun affecting the amount of sunlight arriving at certain latitudes, initiating changes to the carbon cycle.[10] Atmospheric CO_2 and other greenhouse gases

---

[5] Fn.1 (Arrhenius) 268–73.

[6] Spencer Weart, *The Discovery of Global Warming* (Harvard University Press, 2008), chapter 1; also S. R. Weart, 'The Carbon Dioxide Greenhouse Effect' on 'The Discovery of Global Warming' website, American Institute of Physics, February 2018, <http://www.aip.org/history/climate> accessed 30/01/19.

[7] Ibid (both titles). Although more potent greenhouse gases have since been identified, such as methane, CH_4.

[8] Ibid. The 'Keeling curve' is named after Charles David Keeling who was largely responsible for the CO_2 data available today.

[9] Ibid.

[10] Ibid.

have an amplifying effect on the feedback loops that make up the climate system.[11] Part of the problem is that the dangerous climate changes due to these amplifications – rising sea-levels, retreating glaciers and polar ice caps, loss of the Greenland ice sheet, more extreme heat waves and droughts, more excessive floods and more powerful storms – are not only taking the climate system into unknown territory, but because of the in-built time lag for changes in that system, will continue to be experienced centuries after emissions are reduced: 'The risk of abrupt or irreversible changes increases with the magnitude of the warming'.[12]

In October 2018, the Intergovernmental Panel on Climate Change (IPCC) presented key findings from its assessment of the available scientific, technical, and socio-economic literature relevant to global warming of 1.5°C and for the comparison between global warming of 1.5°C and 2°C above pre-industrial levels.[13] Climate models predict robust differences in regional climate characteristics between the present and warming of 1.5°C, and between 1.5°C and 2°C.[14] Modelling indicates that for pathways with no or limited overshoot of 1.5°C, global net anthropogenic CO_2 must decline by 45 per cent from 2010 levels by 2030 and reach net zero by about 2050.[15]

POLICY RESPONSES TO ADDRESS DANGEROUS CLIMATE CHANGE

1.   Intergovernmental Responses

Identification of climate change as an urgent world problem at an intergovernmental level can be traced to the First World Climate Conference in 1979, which called 'on governments to anticipate and guard against potential climate hazards'.[16] International recognition in the 1980s of the potential problem developing due to growth in anthropogenic emissions causing dangerous climate-altering impacts is reflected in numerous resolutions of the United Nations General Assembly (UNGA). For example, in 1987, it was agreed that

---

[11]   Ibid.
[12]   Intergovernmental Panel on Climate Change (IPCC), Fifth Assessment Report, Climate Change 2014 Synthesis Report, Summary for Policymakers, 73 <http://www.ipcc.ch/pdf/assessment-report/ar5/syr/AR5_SYR_FINAL_SPM.pdf> accessed 20/04/17.
[13]   Intergovernmental Panel on Climate Change (IPCC), *Global warming of 1.5°C*, Special Report, 2018, <https://www.ipcc.ch/report/sr15/> accessed 08/10/18.
[14]   Ibid, B.1.
[15]   Ibid, C.1.
[16]   United Nations Framework Convention on Climate Change: Handbook, 2006, Bonn, Germany: Climate Change Secretariat, 17.

the United Nations Environment Programme (UNEP) 'should attach impor-tance to the problem of global climate change' and cooperate 'closely with the World Meteorological Organization (WMO) and the International Council of Scientific Unions'.[17] Then, in 1988, the General Assembly,[18] reaffirming its earlier resolution,[19] endorsed the WMO and UNEP jointly establishing the IPCC '... to provide internationally coordinated scientific assessments of the magnitude, timing and potential environmental and socio-economic impact of climate change and realistic response strategies ...'[20]

To this end, in its fifth Assessment Report, the IPCC states that '... human influence on the climate system is clear, and recent anthropogenic emissions of greenhouse gases are the highest in history ... Warming of the climate system is unequivocal ...'[21] As to the causes, it is equally unequivocal: 'Anthropogenic greenhouse gas emissions have increased since the pre-industrial era, driven largely by economic and population growth, and are now higher than ever ... and are extremely likely to have been the dominant cause of the observed warming since the mid-20th century'.[22]

The UNGA supported the UNEP and WMO request in 1989 to 'begin prepa-rations for negotiations on a framework convention on climate'.[23] In 1990, a single intergovernmental negotiating process was established under the auspices of the UNGA, for the preparation of 'an effective framework conven-tion on climate change'[24] and the United Nations Conference on Environment and Development (UNCED) in Rio de Janeiro adopted the United Nations Framework Convention on Climate Change (UNFCCC)[25] on 9 May 1992.[26]

---

[17]   UNGA A/RES/42/184, 11 December 1987 <http://www.un.org/en/ga/search/view_doc.asp?symbol=A/RES/42/184> accessed 05/06/17.
[18]   UNGA A/RES/43/53, 6 December 1988 <http://www.un.org/en/ga/search/view_doc.asp?symbol=A/RES/43/53> accessed 05/06/17.
[19]   Ibid, paragraph 3.
[20]   Ibid, paragraph 6.
[21]   Intergovernmental Panel on Climate Change (IPCC), Fifth Assessment Report, Climate Change 2014 Synthesis Report, Summary for Policymakers, 2 <http://www.ipcc.ch/pdf/assessment-report/ar5/syr/AR5_SYR_FINAL_SPM.pdf> accessed 20/04/17.
[22]   Ibid, 4.
[23]   UNGA A/RES/44/207, 22 December 1989, paragraph 10 <http://www.un.org/en/ga/search/view_doc.asp?symbol=A/RES/44/207> accessed 05/06/17.
[24]   UNGA A/RES/45/212, 21 December 1990, paragraph 1 <http://www.un.org/en/ga/search/view_doc.asp?symbol=A/RES/45/212> accessed 05/06/17.
[25]   United Nations Framework Convention on Climate Change, 9 May 1992, 1771 UNTS 107.
[26]   UNFCCC website <http://unfccc.int/essential_background/convention/status_of_ratification/items/2631.php> accessed 05/06/17.

2.   Institutions and Other Bodies

The UNFCCC provides for a number of institutions and other bodies, with the Conference of Parties (COP) as the supreme body of the Convention.[27] The COP is responsible for reviewing implementation of the Convention and any related legal instruments (for instance, the Kyoto Protocol and the Paris Agreement) and for making decisions necessary to promote the effectiveness of that implementation and to that end, a list of roles and functions are set out.[28] A secretariat is established and its functions elaborated.[29] The secretariat's role includes making arrangements for the COP and other Convention bodies, assisting parties, supporting negotiations, and coordinating with secretariats of other relevant international organizations, for instance, the Global Environment Facility (GEF).[30] It also has specific responsibilities including, for example, maintaining the international transaction log under the Kyoto Protocol.

There is provision in the Convention also for two subsidiary bodies to the COP, the subsidiary body for scientific and technological advice (SBSTA)[31] and the subsidiary body for implementation (SBI)[32] and their respective roles set out. The role of the SBSTA is considered more later, in relation to operationalization of the Paris Agreement. The UNFCCC provides also for a financial mechanism, the operation of which is to be entrusted to one or more existing international bodies.[33] The GEF was mandated to be this entity on an interim basis,[34] since formalized pursuant to subsequent decisions of the COP and understandings entered with the GEF.[35] The COP can also establish interim and ad hoc bodies to undertake specific tasks.[36]

---

[27] Article 7(2) UNFCCC.
[28] Ibid, (a)–(m).
[29] Article 8 UNFCCC.
[30] The GEF was established in 1991 by the World Bank, UNEP and United Nations Development Programme (UNDP) to fund projects in developing countries that provide global environmental benefits <http://www.gef.org/>.
[31] Article 9.
[32] Article 10.
[33] Article 11(1).
[34] Article 21(3).
[35] Fn.16 (UNFCCC Handbook) 117.
[36] Article 7(2)(i). Examples include, for instance, the Ad hoc Group on the Berlin Mandate (AGBM).

3.   Principles and Measures

Parties, in acting to achieve the objective of the Convention and implementing its provisions, are to be guided by the principles set out in Article 3, UNFCCC. These include the principles of inter- and intra-generational equity and of common but differentiated responsibilities and respective capabilities;[37] the precautionary principle and need for cost effectiveness in policies and measures;[38] promotion of sustainable development;[39] and that measures should not constitute a means of discrimination or a disguised restriction on free trade.[40] For the purpose of matters addressed by this book, the principle of common but differentiated responsibilities and respective capabilities is of primary interest: developed country parties are called on to take the lead in combating climate change and the principle carries through not only in the chapeau to Article 4 and the differentiated commitments of developed and developing country parties, but through the measures elaborated in the Convention, notably in relation to mitigation. This is most apparent in the Kyoto Protocol (discussed below) and notwithstanding the fundamentally changed approach to mitigation obligations taken in it, the Paris Agreement is required to be implemented to reflect, inter alia, the principle of common but differentiated responsibilities and respective capabilities, in the light of different national circumstances.[41] Differentiation between developed and developing country parties has remained a source of friction in negotiations for the Paris Agreement's operationalization.[42]

(i)   Mitigation

The call from the First World Climate Conference in 1979 had been for governments to anticipate and guard against potential climate hazards, suggesting a focus on adaptation. However, the emphasis in the UNFCCC is clearly on mitigation.[43] The ultimate objective of the UNFCCC and any related legal

---

[37]   Article 3(1).
[38]   Article 3(3).
[39]   Article 3(4).
[40]   Article 3(5).
[41]   Article 2(2) Paris Agreement.
[42]   Daniel Bodansky and Lavanya Rajamani, 'The Issues that Never Die' (2018) 12(3) *CCLR* 184, 189. As to the distinction between the Kyoto Protocol and Paris Agreement in relation to this principle, see also Chapter 4 following.
[43]   Fn.16 (UNFCCC Handbook) 17, 74: 'Mitigating climate change and its impact lies at the heart of the Convention's objective'; 'The main concern of the Convention is clearly mitigation; adaptation has been widely seen as the "poor relation"': David Freestone, 'The United Nations Framework Convention on Climate Change – the Basis for the Climate Change Regime', in C. Carlarne, K. Gray, and R. Tarasofsky (eds.), *Oxford Handbook of International Climate Change Law* (Oxford University Press,

instruments that the COP may adopt is set out as being '… to achieve, in accordance with the relevant provisions of the Convention, stabilization of greenhouse gas concentrations in the atmosphere at a level that would prevent dangerous anthropogenic interference with the climate system'.[44] All parties undertake commitments to '… formulate, implement, publish and regularly update national and, where appropriate, regional programmes containing measures to mitigate climate change by addressing anthropogenic emissions by sources and removals by sinks of all greenhouse gases not controlled by the Montreal Protocol'.[45]

Parties listed in Annex I, broadly the developed country parties, each also undertake more specific commitments to '… adopt national policies and take corresponding measures on the mitigation of climate change, by limiting its anthropogenic emissions of greenhouse gases and protecting and enhancing its greenhouse gas sinks and reservoirs'.[46] These parties undertake further, in order to promote progress, to communicate periodically detailed information on their policies and measures, as well as on resulting projected anthropogenic emissions by sources and removals by sinks of greenhouse gases not controlled by the Montreal Protocol.[47]

(ii)   Adaptation

Notwithstanding this emphasis on mitigation, there are other measures including provisions relating to adaptation, provision of financial resources (mentioned above), developing and transferring technologies, building capacity, and promoting education, training, and public awareness. Adaptation is referenced in a number of articles, including the overall objective that stabilization of GHG emissions should be at such a level that would prevent dangerous anthropogenic interference with the climate system within a timeframe that will allow ecosystems to adapt naturally to climate change.[48] All parties are to implement measures, inter alia, to facilitate adequate adaptation to climate change;[49] cooperate in preparing for adaptation to the impacts of climate change;[50] and employ appropriate methods to minimize adverse effects on

---

2016) 100; also Daniel Bodansky, Jutta Brunnée, and Lavanya Rajamani, *International Climate Change Law* (1st edn., Oxford University Press, 2017) 12–13.

[44]   Article 2 UNFCCC.
[45]   Article 4(1)(b).
[46]   Article 4(2)(a).
[47]   Article 4(2)(b). The parties to the Paris Agreement have undertaken more substantial commitments in this respect, for example, in Articles 4 and 13.
[48]   Fn.44 (Art. 2).
[49]   Fn.45 (Art. 4(1)(b)).
[50]   Article 4(1)(e).

the economy, public health, or the environment from projects or measures to (mitigate or) adapt to climate change.[51] Annex II-listed parties (developed countries, excluding economies in transition) are required to assist particularly vulnerable developing countries to meet the costs of adaptation.[52]

(iii)   Other measures

All parties must develop, update, and publish national inventories of anthropogenic GHG emissions by sources and removals by sinks.[53] This must be in accordance with Article 12, which defines national communications in greater detail. Parties are called on also to promote and cooperate in the development, application, and diffusion, including transfer, of technologies that control, reduce, or prevent emissions of GHGs in the relevant sectors, including energy, transport, industry, agriculture, forestry, and waste management.[54] Annex II-listed parties are called on to provide financial resources for technology transfer[55] and all developed country parties are urged to take all practical steps to facilitate and finance the transfer of environmentally sound technologies, especially to developing country parties.[56]

Related to technology transfer, capacity building is explicitly addressed in terms of developed countries supporting development and enhancement of endogenous capacities of developing countries.[57] All parties are called on to strengthen systematic observation and scientific and technical capacities[58] and, in doing so, to cooperate in improving developing countries' endogenous capacities.[59]

4.   Further Elaboration of Commitments

Notwithstanding the elements outlined above, as the name indicates, the UNFCCC remains very much a framework agreement and, to address the concerns of some parties, included provision for review of the commitments undertaken by Annex I-listed parties in Articles 4(2)(a) and (b).[60] The first such review was carried out at the first Conference of Parties (COP 1) in Berlin. The

---

[51]   Article 4(1)(f).
[52]   Article 4(4).
[53]   Article 4(1)(a). See also Article 13, Paris Agreement.
[54]   Article 4(1)(c).
[55]   Article 4(3).
[56]   Article 4(5).
[57]   Ibid.
[58]   Articles 4(1)(g), 5(a) and (b).
[59]   Article 5(c).
[60]   Article 4(2)(d).

decision resulting from this review,[61] known as the 'Berlin Mandate', agreed to strengthen the commitments of Annex I-listed parties. Pursuant to the process to review Article 4(2)(a) and (b), in carrying out the Berlin Mandate, the COP decided, at its third meeting, to adopt the Kyoto Protocol.[62]

The strengthening of commitments, referred to in the Berlin Mandate, translated into agreement by UNFCCC Annex I-listed parties, that were also parties to the Kyoto Protocol (the Annex B Parties), to be bound by specific commitments on GHG limitation or reduction, as listed in Annex B of the Protocol.[63] The Protocol also contains a range of provisions for flexibility, including three so-called flexibility mechanisms, being joint implementation,[64] the clean development mechanism,[65] and international emissions trading.[66] Each Annex B Party's commitment, expressed as its assigned amount,[67] was divided into assigned amount units (AAUs), defined in the modalities, rules, and guidelines for emissions trading under Article 17,[68] as being '… equal to one metric tonne of carbon dioxide equivalent, calculated using global warming potentials defined by decision 2/CP.3 or as subsequently revised in accordance with Article 5'.[69] Also set out in that decision are the modalities for the accounting of assigned amounts under Article 7, paragraph 4 of the Protocol.[70] Paragraph 13 of that decision provides 'Each Party included in Annex I shall retire ERUs, CERs, AAUs and/or RMUs for the purpose of demonstrating its compliance with its commitment under Article 3, paragraph 1'.[71]

---

[61]   Decision 1/CP.1, FCCC/CP/1995/7/Add.1 <http://unfccc.int/resource/docs/cop1/07a01.pdf> accessed 05/06/17.

[62]   Decision 1/CP.3, FCCC/CP/1997/7/Add.1 <http://unfccc.int/resource/docs/cop3/07a01.pdf> accessed 05/06/17.

[63]   Article 3, Kyoto Protocol to United Nations Framework Convention on Climate Change, 11 December 1997, 2303UNTS162 (2005).

[64]   Article 6.

[65]   Article 12.

[66]   Article 17. The principal focus of this book is on international emissions trading (IET), although, for these purposes consideration also needs to include how the clean development mechanism (CDM) has performed. This is necessary due to the significant volume of trading in CDM project-generated certified emission reductions, as part of the IET market – far more than for the other tradable units created.

[67]   Calculated as per Article 3(7), Kyoto Protocol.

[68]   Decision 11/CMP.1, FCCC/KP/CMP/2005/8/Add.2, 17 <http://unfccc.int/resource/docs/2005/cmp1/eng/08a02.pdf#page=17> accessed 06/06/17.

[69]   Ibid at paragraph 3; similar definitions are set out for the other tradable units, certified emission reductions (CERs) under the Clean Development Mechanism, emission reduction units (ERUs) under joint implementation and removal units (RMUs) from land use, land-use change and forestry activities (LULUCF).

[70]   Decision 13/CMP.1, FCCC/KP/CMP/2005/8/Add.2, 23 <http://unfccc.int/resource/docs/2005/cmp1/eng/08a02.pdf#page=23> accessed 06/06/17.

[71]   Ibid, 27.

Emissions trading, or tradable permit schemes, have been implemented 'to deal with various environmental or resource problems since the 1970s', including different types of air pollution, fisheries management, water management, waste management, and land-use.[72] While these earlier illustrations seem to have been successful,[73] there are fundamental differences between these schemes and an international emissions trading scheme. For instance, the earlier schemes being domestic, the market mechanism would function as part of a licensing regime (thus, invoking compliance and potential enforcement) for participating entities. Furthermore, in an operational sense, schemes such as the US Environmental Protection Agency's (USEPA) Acid Rain Program[74] benefitted from real-time monitoring of (smoke stack) emissions,[75] which is not possible for international carbon emissions trading.

Nevertheless, the basic underlying logic for emissions trading remains valid, relying on the 'fundamental tenet [of] the exploitation of cost heterogeneity to minimise overall compliance costs',[76] or as explained in Chapter 1, achieving emission reductions in the most cost efficient location first. Mitigation of GHG emissions means changing behaviour across a range of vectors, thus changing the way many economic activities are carried out and so, it is assumed, their effects. Imposing a climate change related price on carbon emissions is one way to do this, in conjunction with environmental regulation, creating a market

---

[72] Cédric Philibert and Julia Reinaud, 'Emissions Trading: Taking Stock and Looking Forward' OECD Environment Directorate/International Energy Agency, COM/ENV/EPOC/IEA/SL T(2004)3, 8.

[73] See, for instance, Tom Tietenberg et al., *International Rules for Greenhouse Gas Emissions Trading, Defining the principles, modalities, rules and guidelines for verification, reporting and accountability*, (1999), UNCTAD, 6 <UNCTAD/GDS/ MDP/G24/2008/4 - gdsmdpg2420084_en.pdf> accessed 10/05/17; A. Ellerman et al., 'Emissions Trading in the U.S., Experience, Lessons and Considerations for Greenhouse Gases', May, 2003, Pew Center on Global Climate Change <https://www .c2es.org/publications/emissions-trading-us-experience-lessons-and-considerations -greenhouse-gases> accessed 09/05/17.

[74] 42 U.S.C. United States Code, 2011 Edition, Title 42 - The Public Health and Welfare, Chapter 85 - Air Pollution Prevention and Control, Subchapter IV-A - Acid Deposition Control <https://www.gpo.gov/fdsys/pkg/USCODE-2011-title42/html/ USCODE-2011-title42-chap85-subchapIV-A-sec7651b.htm> accessed 06/07/17.

[75] Known as CEMS – the continuous emissions monitoring system, it added about 7 per cent to total compliance costs, whereas materials balancing could have produced equally accurate estimates at less cost: Fn.73 (Ellerman et al.) 17.

[76] Shi-Ling Hsu, 'International Market Mechanisms', in C. Carlarne, K. Gray, and R. Tarasofsky (eds.), *Oxford Handbook of International Climate Change Law* (Oxford University Press, 2016) 241, citing William J. Baumol and Wallace E. Oates, *The Theory of Environmental Policy* (2nd edn, Cambridge University Press, 1988) 21–3 and Tom Tietenberg and Lynne Lewis, *Environmental and Natural Resource Economics* (10th edn, Pearson, 2014) 357.

through which the environmental cost of those carbon emissions becomes internalized in the relevant economic activities.[77] The larger the market, the greater the efficiency benefits it might be expected to yield.[78]

It is not the purpose of, nor essential to the proposals outlined in, this book to explore the economic reasoning that underpins emissions trading in greater detail than already set out. Rather, this book proceeds on the basis that economic theory supporting emissions trading is settled, at least to the extent that emissions trading has become the mitigation policy measure of choice, adopted by many jurisdictions.[79] The focus in considering implementation of emissions trading as a policy mechanism is on IET, pursuant to Article 17 of the Kyoto Protocol. The carbon market that developed through IET and is considered here consists of trading in AAUs, certified emission reductions (CERs) pursuant to Article 12 Kyoto Protocol and EU Allowances (EUAs) pursuant to the European Union Emission Trading Scheme (EUETS).

5.   Issues Encountered with the Kyoto Protocol

In 2014, the IPCC observed:

> ... mechanisms that set a carbon price ... have been implemented with diverse effects due in part to national circumstances as well as policy design. The short-run environmental effects of cap and trade systems have been limited as a result of loose caps or caps that have not proved to be constraining ...[80]

The Kyoto Protocol's flexible mechanisms were intended to make the task of mitigating global GHG emissions more cost effective and economically efficient. Derived from a process of negotiation between 197 parties, however,

---

[77]   Justin Macinante, 'A Conceptual Model for Networking of Carbon Markets on Distributed Ledger Technology Architecture' [2017] *CCLR* 243. See also fn.72 (Philibert and Reinaud (2004)) reviewing the origins of tradable permit schemes and emissions trading; also see Richard Schmalensee and Robert N. Stavins, 'Lessons Learned from Three Decades of Experience with Cap-and-Trade' Discussion Paper 2015-80. Cambridge, Mass.: Harvard Project on Climate Agreements, November 2015.

[78]   Fn.76 (Hsu).

[79]   Richard Baron and Cedric Philibert, 'Act Locally, Trade Globally Emissions Trading for Climate Policy', © OECD/IEA, 2005, 22 <https://www.iea.org/publications/freepublications/publication/act_locally.pdf> accessed 14/05/17.

[80]   Intergovernmental Panel on Climate Change (IPCC), Climate Change 2014: Synthesis Report. Contribution of Working Groups I, II and III to the Fifth Assessment Report of the Intergovernmental Panel on Climate Change, 4.4.2.2 Mitigation, 107 [Core Writing Team, R.K. Pachauri and L.A. Meyer (eds.)]. IPCC, Geneva, Switzerland.

it was probably inevitable that there would be issues with the design outcome and, in particular, political issues.

In a political sense, the fate of the Kyoto Protocol flexible mechanisms was tied to some extent to the domestic situation in the US. At the time of the Third Conference of Parties to the UNFCCC (COP-3), the US was still the largest emitter, accounting for approximately 15.5 per cent of global emissions of carbon dioxide equivalent (CO_2e) GHG emissions, with the European Union at 12.26 per cent and the People's Republic of China (PRC) not far behind at 11.79 per cent.[81] However, PRC emissions were rapidly increasing and, in other respects, the dynamics of the respective emissions were changing. At the time of the Kyoto Protocol signing, 'EU emissions remained at 1990 volumes, while North America's had grown 14 per cent, and those of Russia and Ukraine had dropped 30 per cent'.[82]

As outlined at the Eighth Meeting of the Ad Hoc Group on the Berlin Mandate (AGBM-8) just weeks before COP-3, the US was clear that it would not assume binding obligations unless key developing countries participated meaningfully.[83] In response, the negotiating group of 77 developing countries with China (G-77/China) 'used every opportunity to distance itself from attempts to draw developing countries into … new commitments'.[84] Twelve months later, at the corresponding meeting of the Subsidiary Bodies prior to COP-4 in Buenos Aires, the issue remained unresolved[85] and COP-4 itself, in November 1998, provided the backdrop to a dramatic development, when US President Clinton signed the Kyoto Protocol despite strong domestic political opposition.[86] Then in March 2001, the new Bush Administration 'declared its opposition to the Protocol, stating that it believed it to be "fatally flawed", as

---

[81]   In 1997, the US emissions were 6,724,414.4 kt CO_2e gas; the EU emissions were 5,318,851.567 kt CO_2e gas; and PRC emissions were 5,113,706,854 kt CO_2-eq gas, out of the world total of 43,375,207.968 kt CO_2e gas: World Bank data <http://data .worldbank.org/indicator/EN.ATM.GHGT.KT.CE> accessed 15/06/17.

[82]   Fn.79 (Baron, Philibert) 43.

[83]   International Institute for Sustainable Development, Earth Negotiations Bulletin, Vol.12, No.66, Report of the Meeting of the Subsidiary Bodies to the FCCC: 21–30 October 1997, 3 November 1997, 3 <http://enb.iisd.org/download/pdf/enb1266e.pdf> accessed 06/06/17.

[84]   Ibid, 16.

[85]   International Institute for Sustainable Development, Earth Negotiations Bulletin, Vol.12, No.86, Report of the Meetings of the FCCC Subsidiary Bodies: 2–12 June 1998, International Institute for Sustainable Development, 15 June 1998, 12 <http://enb .iisd.org/download/pdf/enb1286e.pdf > accessed 06/06/17.

[86]   International Institute for Sustainable Development, Earth Negotiations Bulletin, Vol.12, No.97, Report of the Fourth Conference of the Parties to the United Nations Framework Convention on Climate Change: 2–13 November 1998, 13 December 1997, 13 <http://enb.iisd.org/download/pdf/enb1297e.pdf> accessed 06/06/17.

it would damage its economy and exempted developing countries from fully participating'.[87] The US withdrawal, after the failed COP-6 at The Hague, meant that to take effect, the Protocol became dependent on ratification by the Russian Federation.[88]

This left the EU and 37 Annex B Parties, of which 13 were economies in transition (EITs), with commitments, out of 197 UNFCCC signatories. Taking OECD countries as a proxy for the Annex B Parties (thus excluding EITs), with the US included, would still have represented only approximately 37.95 per cent of 1997 global emissions. Without the US, they represented only about 22.45 per cent of global emissions,[89] and as noted earlier, of this amount, the EU represented 12.26 per cent. Hence, the limited extent of coverage for international emission trading raised questions whether, from the outset, the regime could have had any significant impact in mitigating global emissions.

Two other problems encountered were the inflexibility of the Kyoto Protocol, a reflection on the treaty process itself, and the lack of support, in the end, that it received. Demonstrative of inflexibility were attempts made by Kazakhstan, beginning at COP-4, to become an Annex B Party and voluntarily to take on QELRCs, so that it might engage in emissions trading and thereby gain access to financing for low-carbon development.[90] Ten COPs and a decade after initiating the process, Kazakhstan would still require a 75 per cent majority vote and even then, the change would only enter into force for those state parties that were to ratify the amendment.[91] It is difficult to disagree with the description of this situation as a 'dysfunction' of the UNFCCC monopoly, an inflexibility that does not encourage much additional effort by governments.[92]

The lack of support the Kyoto Protocol has received is epitomized, not only in terms of the direct physical consequences, but also perhaps more significantly in terms of the political message conveyed, by the withdrawal of the

---

[87]   International Institute for Sustainable Development, Earth Negotiations Bulletin, Vol.12, No.176, Summary of the Resumed Sixth Session of the Conference of Parties to the UN Framework Convention on Climate Change: 16–27 July 2001, 30 July 2001, 2 <http://enb.iisd.org/download/pdf/enb12176e.pdf> accessed 06/06/17.

[88]   Ibid, 13.

[89]   World Bank data <http://data.worldbank.org/indicator/EN.ATM.GHGT.KT .CE> accessed 15/06/17.

[90]   Annie Petsonk, 'Docking Stations: Designing a More Open Legal and Policy Architecture for A Post-2012 Framework to Combat Climate Change' (2009) 19(3) *Duke Journal of Comparative and International Law* 433, 441–3; also Robert O. Keohane and David G. Victor, 'The Regime Complex for Climate Change' (2011) 9(1) *Perspectives on Politics* 7, 15.

[91]   Ibid (Petsonk) 443.

[92]   Fn.90 (Keohane and Victor) 15.

US in 2001. In an ongoing sense, it is evidenced most notably in the absence of state parties willing to sign up for the second commitment period (CP.2), which covers only 12 per cent of global GHG emissions. As of May 2014, only nine countries had ratified the Doha Amendment to the Kyoto Protocol by which CP.2 was agreed, a long way short of the 144 countries required.[93] By January 2020, 136 of the required 144 letters of acceptance had been deposited, still insufficient for it to take effect.[94]

Notwithstanding these considerations, all of the Annex B countries that fully participated in the first commitment period (CP.1) complied with their commitments and only nine needed to resort to the flexible mechanisms in order to do so.[95] Shishlov and colleagues ascribe this overachievement to four factors: hot air from the EITs (see next section); non-participation of the US and Canada; the global financial and economic crisis; and policies and measures put in place by the signatory countries. While hardly a ringing endorsement, there are both positive and negative lessons,[96] one of these being that 'compliance does not of itself equate to impact',[97] highlighting a point about mitigation policy architecture, that '... trade-offs between breadth of participation and depth of commitments are central to the design of international instruments'.[98] However, the evidence seems to indicate that while there was compliance with the CP.1 commitments, the Kyoto Protocol has not achieved either breadth or depth in terms of outcome.

OPERATIONAL IMPLEMENTATION

1. What Reductions Do Allowances Represent?

The issue referred to as hot air is an early indicator of the fundamental conceptual problem of the nature of what is traded.[99] Initially considered a red line for the EU, as negotiations developed the hot air issue seemed to recede

---

[93] World Bank, Ecofys, 2014, *State and Trends of Carbon Pricing 2014*, Washington, DC: World Bank, 14.
[94] UNFCCC website <https://unfccc.int/process/the-kyoto-protocol/the-doha-amendment> accessed 22/01/20.
[95] Igor Shishlov, Romain More, and Valentin Bellassen, 'Compliance of the Parties to the Kyoto Protocol in the First Commitment Period' (2016) 16(6) *Climate Policy* 768, 770.
[96] Michael Grubb, 'Full Legal Compliance with the Kyoto Protocol's First Commitment Period – Some Lessons' (2016) 16(6) *Climate Policy* 673.
[97] Ibid, 674.
[98] Fn.43 (Bodansky, Brunnée, Rajamani) 26.
[99] Considered in Chapter 4 and later chapters.

in significance for IET. Nonetheless, it provides important background to the issues considered later in this book.

Former Iron Curtain states, whose economies collapsed in the late 1990s, included states that had agreed to be bound by specific commitments on GHG limitation or reduction (economies in transition, EITs), as listed in Annex B of the Kyoto Protocol.[100] The commitments, in the case of these now collapsed economies, far exceeded their actual emission levels. As a result, the quantified emission limitation and reduction obligations (QELROs) for these Annex B Parties, and hence the assigned amount for each calculated on the basis of its QELRO, far exceeded their requirements for compliance over the commitment period,[101] thereby providing them with immediate surpluses of AAUs that they could sell. Thus 'At AGBM 7 … the problem that became dubbed "hot air" was first raised. The EU expressed concern that the buying up of past emission reductions, notably in EITs, could mean that commitments were fulfilled without further emission reductions being achieved'.[102]

This issue would have been more problematic had there been significant transfers of AAUs, but there were not, purchasers being limited to only a few countries. Data compiled by UNEP indicates that the main country purchasers of AAUs were Japan (mainly through firms with domestic commitments), Spain, and Austria, while the World Bank was also a purchaser (for its funds). The main sellers were Czech Republic, Estonia, Hungary, Latvia, and Poland.[103]

Nevertheless, this issue demonstrates a structural shortcoming in the design of IET that goes to the credibility of what it seeks to achieve. Additionally, while AAUs could not be used for compliance by installations with obligations under the EUETS, EU countries would be able use AAUs to fulfil their obli-

---

[100] This paragraph draws on the author's previous publication 'From Homogeneity to Heterogeneity and the Fundamental Question – What is Being Traded?' University of Edinburgh School of Law Research Paper Series No.2017/15 <https://ssrn.com/abstract=3015798>.

[101] Pursuant to Article 3, Kyoto Protocol.

[102] Joanna Depledge, 'Tracing the Origins of the Kyoto Protocol: An Article-by-Article Textual History', Technical Paper, FCCC/TP/2000/2, 25 November 2000, prepared under contract to UNFCCC, August 1999/August 2000, UNFCCC, 83. This is somewhat ironic, given the current situation of the EUETS. AGBM7 took place 28 July–7 August 1997.

[103] Matthew Ranson and Robert N. Stavins, 'Linkage of Greenhouse Gas Emissions Trading Systems: Learning from Experience'. Discussion Paper ES 2013-2. Cambridge, Mass.: Harvard Project on Climate Agreements, November 2013 in Table 2, based on data from UNEP Risø Centre (2013), 23.

gations more generally.[104] Further, provided the Japanese government agreed, Japanese companies were able to purchase AAUs to meet commitments under the domestic Japanese voluntary scheme,[105] private firms (intermediaries) generally could purchase AAUs, as did World Bank funds.[106] As these transactions were usually bilateral, they are not recorded on any public exchange, so price and other market information would not be readily accessible.[107] AAU trading is tracked by the International Transaction Log (ITL), but while this is hosted by the UN and publicly available, the ITL only records transactions between countries.[108]

2.   Clean Development Mechanism Implementation

The Clean Development Mechanism (CDM) has been the most successful tool for mobilizing mitigation finance in less developed countries, accounting for most of the project-based market activity.[109] All the same, at the commencement of the Kyoto Protocol first commitment period in March 2008, it was reported that the CDM was suffering from procedural inefficiencies and regulatory bottlenecks.[110] These included a substantial backlog of projects awaiting validation; market participants taking up to six months to engage a Designated Operational Entity (DOE) due to lack of capacity in the market; an average delay of 80 days for projects to become registered; and projects taking between one and two years in the pipeline before achieving their first issuance.[111]

Despite the difficulties, at its peak in 2012 over 3400 projects and Programs of Activities (PoAs) (an average of over nine per day) were registered and approximately 339 Mt CO_2-eq GHGs reduced.[112] In total, over the ten years of operation to 2014, more than 7,700 projects and PoAs were registered

---

[104] Elizabeth L. Aldrich and Cassandra L. Koerner, 'Unveiling Assigned Amount Unit (AAU) Trades: Current Market Impacts and Prospects for the Future' (2012) 3 *Atmosphere* 229–45.

[105] Ibid.

[106] Ibid, 232.

[107] Ibid.

[108] Ibid, 233.

[109] In 2007, 87 per cent by volume, 91 per cent by value transacted: Jolene Lin, 'Private Actors in International and Domestic Emissions Trading Schemes' in David Freestone and Charlotte Streck (eds.), *Legal Aspects of Carbon Trading: Kyoto, Copenhagen and Beyond* (Oxford University Press, 2009) 139, citing World Bank, *State and Trends of the Carbon Market 2008* (Washington, DC, May 2008) 19.

[110] Ibid (World Bank) 4.

[111] Ibid.

[112] World Bank, Ecofys, *State and Trends of Carbon Pricing 2014* (Washington, DC, 2014) 45.

and US$130 billion invested in GHG reducing activities.[113] All the same, the CDM has been beset by numerous design issues and has been in a constantly evolutionary state. In broad terms, these issues have included the complexity of its procedures, leading to delays and high transaction costs; registration of ineligible (non-additional) projects; projects (especially industrial gases, for example, elimination of HFC-23) which gamed the system; and projects being concentrated in particular countries, especially China and India, which benefited from the majority of projects.[114]

3.   Operationalization of the European Carbon Market

It is not intended to drill down into the experience of domestic (or regional) emissions trading schemes generally, however, the EUETS warrants individual consideration simply because it has been, by far, the largest component of what might be described as the global carbon market. As such, it can be seen as illustrative of how the global carbon market has operated since its inception.

The EUETS has grown '… from what was initially perceived as a "policy experiment" into the flagship of EU climate policy'.[115] This is despite the fact that, originally, the European Commission (EC) had favoured a carbon tax rather than emissions trading.[116] The scheme's growth has not been without problems, commencing with the decentralized system that evolved in which crucial policy aspects were left to individual member states.[117] In the case of cap setting for the first phase of the scheme, for instance, the outcome was that various member states over-allocated. The 'uncoordinated leak and release of verified emissions', in April 2006, revealing this over-allocation, caused the EUA price (and, consequently, the CER price) to plummet.[118]

Since that time there have been moves to 'a more centralized and harmonized approach',[119] however, over-supply issues continue to impact the EUETS. A combination of the economic downturn, compounded by an excess

---

[113]   Ibid.

[114]   Ibid. The percentages of registered projects by host country show China 50 per cent and India 20 per cent (UNFCCC, UNEP Risø Centre).

[115]   Markus Pohlmann, 'The European Union Emissions Trading Scheme' in David Freestone and Charlotte Streck (eds.), *Legal Aspects of Carbon Trading: Kyoto, Copenhagen and Beyond* (Oxford University Press, 2009) 339.

[116]   Marjan Peeters, 'Greenhouse Gas Emissions Trading in the EU' in D.A. Farber and M. Peeters (eds.), *Encyclopedia of Environmental Law: Volume 1 Climate Change Law* (Edward Elgar, 2016) 378.

[117]   Fn.115 (Pohlmann) 343.

[118]   Ibid, 354; World Bank, *State and Trends of Carbon Pricing 2007* (Washington, DC, 2007) 12.

[119]   Fn.115 (Pohlmann) 344; also fn.116 (Peeters) 381.

of CERs being available, led to a significant surplus in the scheme, which has depressed prices.[120] Rule changes relating to the use of CERs and 'back-loading' to postpone the auction of 900 million EUAs, have stabilized prices. The introduction of the market stability reserve allows greater flexibility in managing the market.[121]

Apart from these structural issues, the EUETS has also suffered operational issues, being a target of VAT fraud and thefts from registries. In August 2009, it was reported that the UK had arrested nine individuals as part of an investigation into a suspected £38 million VAT fraud. The technique usually employed by fraudsters involved purchase and on-sale of emissions allowances, charging VAT to their buyer, but not remitting it to the tax authorities.[122] In September 2009, the EC announced that member states could temporarily apply a reverse charging mechanism, so that VAT liability would be with the seller, while other states applied a zero-rating to EUAs.[123]

The theft of almost four million EUAs, leading to the closure of emissions registries in 2010, was reported as having a major effect on trading.[124] Although 15 of the 30 registries had reopened by March 2011, the matter was complicated by the fact that laws relating to the purchase by a bona fide buyer, in good faith, of stolen goods, vary from one EU country to another. For instance, in the Netherlands and Germany, unwitting buyers would be afforded greater protection than in the UK, with the consequence that some traders began favouring registries in the former states as security against the risk of inadvertently purchasing stolen EUAs,[125] while others were reported to be pulling out of the spot market altogether.[126]

Experts were quoted as ascribing the problem to the fact that the legal nature of EUAs has never been defined,[127] calling for them to be treated in the same

---

[120] World Bank, *State and Trends of Carbon Pricing 2014* (Washington, DC, 2014) 17. The author acknowledges that this generalized statement glosses over the complex set of factors that influences movements in market price in the EUETS. However, over-supply is probably the most influential of those factors.

[121] Ibid, 55.

[122] Carbon Finance, 'EC Proposes VAT Changes to Address Carbon Fraud' (October 2009) 6(10) *Carbon Finance* 8.

[123] Ibid.

[124] Carbon Finance, 'Half of EU's Registries Reopen' (March 2011) 8(2) *Carbon Finance* 1.

[125] Carbon Finance, 'Stolen EUAs Spread Chill Through Carbon Market' (December 2010–January 2011) 7(12) *Carbon Finance* 1-2.

[126] Carbon Finance, 'Commission Disappoints with Response to EUA Scandal' (March 2011) 8(2) *Carbon Finance* 7.

[127] This issue is discussed in detail in Chapter 4.

way as cash, giving the bearer title.[128] However, the problem was seen also as being due, at least in part, to the fact that while derivatives contracts relating to emission allowances were captured by the EU's financial markets regulation, the spot market where the thefts had taken place, was unregulated.[129]

This was one theme, amongst a number, picked up by the European Court of Auditors (ECA), which in 2015 reported on the integrity and implementation of the EUETS.[130] The ECA, noting that emissions markets need sufficient liquidity to function well, suggested the EUETS 'could improve if there were more certainty over an EU-wide definition of allowances, and if allowances were more commercially interesting to voluntary market participants, for example, by supporting the ability to create and protect secure and enforceable security interests' however, the 'EUETS Directive did not define legal status', only explaining ways in which allowances could be used.[131] Other ECA themes included the role of the EUETS as a financial market, emissions derivatives constituting more than 90 per cent of the market, and need for ongoing and effective cooperation between carbon markets and financial markets regulatory bodies.[132] These themes are picked up in following chapters.

---

[128] Fn.124 (Carbon Finance) 2.

[129] Fn.125 (Carbon Finance).

[130] European Court of Auditors, Special Report 'The integrity and implementation of the EU ETS' No.06, 2015 <http://www.eca.europa.eu/Lists/ECADocuments/SR15_06/SR15_06_EN.pdf> accessed 23/06/17.

[131] Ibid, paragraphs 25–28.

[132] Ibid, paragraphs 12–24. As to the more positive outcomes from the EUETS to date, see: Mirabelle Muûls et al., 'Evaluating the EU Emissions Trading System: Take it or Leave it? An Assessment of the Data after Ten Years', October 2016, Imperial College London, Grantham Institute, Briefing Paper No. 21 <https://www.imperial.ac.uk/media/imperial-college/grantham-institute/public/publications/briefing-papers/Evaluating-the-EU-emissions-trading-system_Grantham-BP-21_web.pdf> accessed 16/03/17.

PART II

The carbon market from three perspectives

3. Compartmentalization of the carbon market

International emissions trading (IET) – the carbon market that developed under the Kyoto Protocol – was not designed, but the outcome of international negotiations, fraught with all the compromises such a multilateral process entails. In the outcome, it functions as both a climate policy mitigation measure and as a financial market, operating at an international level. It is posited that for this functional duality to work, there needs to be balance, such that these two functions would be mutually reinforcing. Yet examination of this duality of purposes, it is argued, discloses not balance, but imbalances: first, at a broad, macroscopic level, its evolution resulted in IET, as a financial market, being functionally compartmentalized in the climate policy regime, as a result of which its effectiveness has been impaired (considered in this chapter); while, from a granular, micro-perspective, the focus has been on what is traded and the entitlements attaching thereto, tilting the imbalance the other way, towards the market function (considered in the next chapter). This, also, has detracted from its effectiveness.

This chapter examines the proposition that compartmentalization resulted in impairment of the effectiveness of the carbon market.[1] To better frame the argument being made here, it is proposed to draw on a technique from the field of study of fragmentation in international law. In borrowing this analytical approach, it is appropriate to begin by briefly explaining its genesis, consequently, this chapter proceeds, first, by considering what is fragmentation in international law; second, it examines the 'borrowed' analytical approach, which is normally applied to the interaction between regimes and the consequences of such interactions; then third, that analytical approach is applied as a way of examining the effects of the perceived functional compartmentalization of IET in the climate regime.

---

[1]   See Chapter 1 for elaboration on use of the expression 'carbon market'.

FRAGMENTATION IN INTERNATIONAL ENVIRONMENTAL LAW

The phenomenon of fragmentation in international law derives from the 'specialized and (relatively) autonomous rules and rule-complexes, legal institutions and spheres of legal practice' in international law that have grown up over time in response to social changes.[2] Thus, in a field once seen as being governed by general international law, the operation of specialist systems has developed, and continues to develop, in areas such as trade law, environmental law, human rights law and many others.[3] As a result:

> The problem, as lawyers have seen it, is that such specialized law-making and institution-building tends to take place with relative ignorance of legislative and institutional activities in the adjoining fields and of the general principles and practices of international law. The result is conflicts between rules or rule-systems, deviating institutional practices and, possibly, the loss of an overall perspective on the law.[4]

In its 2006 report, the International Law Commission (ILC) saw these developments in both a positive and a negative light. On one hand, fragmentation reflected 'the rapid expansion of international legal activity into new fields and the diversification of its objects and techniques', while on the other, it created 'the danger of conflicting and incompatible rules, principles, rules-systems and institutional practices'.[5]

The background to concern about fragmentation of international law was 'the rise of specialized rules and rule-systems that have no clear relationship to each other'.[6] This is especially the case with environmental treaties, it has been observed, 'as most matters bear a relationship to the environment'.[7]

---

[2] International Law Commission, *Fragmentation of International Law: Difficulties Arising from the Diversification and Expansion of International Law* Report of the Study Group of the International Law Commission, United Nations General Assembly, A/CN.4/L.682, 13 April 2006, paragraph 8; this report notes at paragraph 5 that 'the background of fragmentation was sketched already half a century ago', referring to an article published in 1953: with the growth in treaty-making over the intervening period, it has clearly become more of an issue. Also see Cinnamon Carlarne, 'International Treaty Fragmentation and Climate Change' in D. A. Farber and M. Peeters (eds.), *Encyclopedia of Environmental Law: Volume 1 Climate Change Law* (Edward Elgar, 2016) 261–72.

[3] Ibid (ILC).

[4] Ibid.

[5] Ibid, paragraph 14.

[6] Ibid, paragraph 483.

[7] Fn.2 (Carlarne) 264, citing ILC (Fn.2 (ILC) paragraph 273, note 358).

International environmental law can be described, on the one hand, 'as a special regime of international law that emerged in response to the growing concern about a number of shared environmental problems', or on the other, as the aggregation of elements of 'public and private international law that are relevant to dealing with environmental challenges'.[8] Whichever of these ways it is perceived, it consists of 'a set of specialized treaties ... which contain diverse environmental objectives and create a series of disparate law-making and implementation organs'.[9]

The ILC report is concerned with techniques for dealing with tensions or conflicts between legal rules and principles, in the substance of international law.[10] It deals, therefore, as a preliminary matter, with the question of what is a conflict,[11] adopting a 'wide notion of conflict as a situation where two rules or principles suggest different ways of dealing with a problem'.[12] However, fragmentation in international law gives rise to a broader suite of issues than just substantive conflict between treaties. For instance, 'it is increasingly evident that not only treaties can be in conflict, but conflicts may also emanate from the decisions of treaty bodies ... international rules on norm conflicts cannot be applied without asking whether – and to what extent – the decisions adopted by these bodies constitute international law-making in the traditional sense ...'[13]

> De jure, however, if the decisions by treaty bodies are not regarded as international lawmaking, the regime of the Vienna Convention on the Law of Treaties (VCLT 1969) does not apply, thus limiting the usefulness of international law in addressing this consequence of fragmentation arising from the climate regime.[14]

One of the conclusions these authors reach is that climate change, reflecting 'an increasingly fragmented body of international environmental norms, poses challenges that urge international lawyers and policymakers to rethink the extent to which international law provides the proper tools to deal with fragmentation.'[15] Climate change law 'is more complex and requires more

---

[8] Ibid, citing P. Birnie, A. Boyle, and C. Redgwell, *International Law and the Environment* (Oxford University Press, 2009).
[9] Ibid.
[10] Fn.2 (ILC) paragraph 21.
[11] Ibid, paragraphs 21–6.
[12] Ibid, paragraph 25.
[13] Harro van Asselt, Francesco Sindico, and Michael Mehling, 'Global Climate Change and the Fragmentation of International Law' (2008) 30(4) *Law & Policy* 423, 430.
[14] Ibid.
[15] Ibid, 440.

complex solutions than merely looking at the relationship between specific treaty regimes'.[16] The nature of this complexity is explored in the next sub-section.

INTERACTION OTHER THAN CONFLICT

It has been argued that, while the ILC report makes an 'in-depth assessment of the difficulties' of fragmentation in international law, it confines itself to analysis of 'the significance of fragmentation for substantive international law', without considering the many 'implications for international institutions and governance structures'.[17] 'Fragmentation means different things to different people' and a variety of examples can be identified in the literature.[18]

Some clarification, van Asselt proposes, could be achieved by: (a) distinguishing between substantive and institutional (where by 'institutions', is meant international judicial bodies) fragmentation; (b) further dividing into 'fragmentation along the lines of issue areas and fragmentation along geographical boundaries'; (c) considering whether the fragmentation referred to 'the relationship between different interpretations of general international law, the relationship between general international law and specialized regimes', or 'among two or more overlapping specialized regimes'; (d) differentiating between fragmentation of primary norms (principal rules of obligation) or that of secondary norms (rules about rules – that is, rules governing the creation, interpretation, and enforcement of primary norms); or (e) fragmentation in terms of sites of 'governance', which broadly takes account of the roles of non-state actors.[19]

After reviewing effects and potential benefits, van Asselt concludes, in this respect, that 'the consequences of fragmentation do not necessarily depend on the existence of overlapping regimes as such, but rather on how their interrelationships are managed'.[20] As this book focuses on a particular measure in the climate regime, and on its interaction, it is helpful to retrace the steps taken by van Asselt in considering types of regime interaction.[21] The first points he makes are that there is a dearth of classifications and typologies in the literature on interactions and that there is confusion over the terms that might be used

---

[16] Fn.2 (Carlarne) 269.
[17] Fn.13 (van Asselt et al) 427.
[18] Harro van Asselt, *The Fragmentation of Global Climate Governance, Consequences and Management of Regime Interactions* (Edward Elgar, 2014) 35.
[19] Ibid, 35–8.
[20] Ibid, 43.
[21] Ibid, ch. 4.

to describe it.[22] While these points are noted, they do not directly impact on the argument propounded here, except that the argument made here might add a further category to any classification devised, being one describing an absence of interaction due to fragmentation, rather than the converse.[23]

Surveying the approaches of scholars to identifying different types of interactions,[24] van Asselt notes a distinction drawn by several scholars between interactions that have 'synergistic, conflicting (disruptive) or neutral/indeterminate effects on the target institution'.[25] However, before considering further the consequences of interactions, van Asselt reviews two other points. First, what is it that 'interacts'? In other words, what is the object of these academic studies? Second, he addresses the need to consider both hard and soft law.

In relation to the first of these points, van Asselt observes that different disciplines have taken different approaches: international relations scholars have 'focused on how institutions and regimes affect each other's development and performance', whereas 'international lawyers have primarily analysed international legal instruments such as treaties'.[26] He notes that the concept of 'regime' has been the object of analysis in studies from both disciplines and that the ILC report frequently uses the term. He adopts a definition that '... regimes are sets of norms, decision-making procedures and organisations coalescing around functional issue-areas and dominated by particular modes of behaviour, assumption and biases'.[27] In relation to the second point, he notes that '[M]ost studies on regime interactions ... focus on traditional, negotiated treaty-based regimes, which constitute hard law', but that 'interactions may also involve regimes not based on legally binding instruments', or soft law.[28]

---

[22]   Ibid, 45: such as 'interactions', 'interlinkages', 'interplay', 'linkages', 'overlap'.

[23]   While the main point being made in this respect in this chapter is that compartmentalization in the case of international emissions trading has caused an absence of interaction that might otherwise have been beneficial to the effectiveness of that mechanism, nevertheless, it is acknowledged that the operation of emissions trading in the climate regime also has the potential to interact in a negative way with other regimes as well, thereby giving rise to a conflict situation. The academic literature notes some instances such as interaction with the international trade regime, however, this issue is not considered as it falls outside the scope of the approach being discussed here.

[24]   Fn.18 (van Asselt) 46–7.

[25]   Ibid, 47, citing Sebastian Oberthür and Thomas Gehring, 'Conceptual Foundations of Institutional Interaction' in Sebastian Oberthür and Thomas Gehring (eds.), *Institutional Interaction in Global Environmental Governance: Synergy and Conflict Among International and EU Policies* (The MIT Press, 2006) 46.

[26]   Ibid.

[27]   Ibid, 49; from Margaret A. Young, 'Introduction: The Productive Friction Between Regimes' in Margaret A. Young (ed.), *Regime Interaction in International Law: Facing Fragmentation* (Cambridge University Press, 2012).

[28]   Ibid, 49.

It is useful to keep in mind the distinction between hard and soft law when considering the approaches that have been taken by policymakers and governments to the application of the market mechanism in international climate law. Even though Article 17 of the Kyoto Protocol sought to introduce international emissions trading as a so-called 'flexible mechanism', it was done in the context of a compliance regime, which included conditions on such trading, for example, in terms of commitment period reserves and the other eligibility requirements.[29] Under Article 6 of the Paris Agreement,[30] parties might have recourse to cooperative approaches involving the international transfer of mitigation outcomes. While the Subsidiary Body for Scientific and Technological Advice (SBSTA) is developing guidance to ensure that robust accounting and the other terms in which Article 6 is expressed are met,[31] unlike the Kyoto Protocol there is no compliance regime sitting over and above the use of these cooperative measures.[32] It might be concluded that, on a continuum with hard law at one end and soft law at the other, international emission trading, to the extent that it takes place under the Paris Agreement, will be more an example of soft law than was the case under the Kyoto Protocol. However, the main point drawn from van Asselt's review, in this context, is that interactions between hard and soft law regimes can take place, and that in global climate governance they are, perhaps, more likely.

Returning to consider the consequences of these interactions, van Asselt distinguishes between conflicting, synergistic, and neutral effects. Addressing conflicts, his first step is to review definitions of what constitutes 'conflict', since it needs to be established at the outset whether it actually exists.[33] In order also to cover policy conflicts, he settles on the 'ILC's broad conceptualisation of conflict "as a situation where two rules or principles suggest different ways of dealing with a problem"', subject to the qualification that those '"different ways" lead to contradictory outcomes'.[34] As to when these interactions could lead to conflicts, van Asselt proposes a list of indicators, comprising: incompatible norms (the clearest such indicator, but also the least likely); diverging objectives; the use of different principles and concepts; opposing economic

---

[29]   Decision 11/CMP.1, FCCC/KP/CMP/2005/8/Add.2, 30 March 2006 <http://unfccc.int/resource/docs/2005/cmp1/eng/08a02.pdf#page=17> accessed 25/04/17.

[30]   FCCC/CP/2015/10/Add.1,   <http://unfccc.int/resource/docs/2015/cop21/eng/10a01.pdf>, accessed 25/04/17.

[31]   Paragraph 36, Decision 1/CP.21, FCCC/CP/2015/10/Add.1, 29 January 2016 <http://unfccc.int/resource/docs/2015/cop21/eng/10a01.pdf> accessed 25/04/17.

[32]   Although note this issue is considered in Chapter 8, where governance structures under the regimes are compared.

[33]   Fn.18 (van Asselt) 52–5.

[34]   Ibid, 53.

incentives; and negative diffusion and learning (that is, lessons being learned that undermine the effectiveness of a regime). He flags also that, apart from a distinction between normative and policy conflicts, a distinction also needs to be drawn between actual and potential conflicts.[35]

The term synergy is considered to mean when the interaction produces a combined effect greater than the sum of the separate effects.[36] However, this leads to the more difficult matter of determining the effectiveness of a regime and, even more so, the aggregate effectiveness of several regimes influencing each other.[37] In the best case, van Asselt states, assessing regime effectiveness is a highly complex task, involving relating 'stated or implicit objectives to observed or anticipated' outcomes, while in the worst case, trying to 'infer any unambiguous regime objective' might be in vain.[38]

Nevertheless, others have sought 'to operationalize "effectiveness" in the context of overlapping regimes' and one such classification, cited by van Asselt, identifies three levels by which a regime's effectiveness can be measured, as: output, being 'norms generated' that correspond to 'cognitive interaction and interaction through commitment'; outcome, being 'behavioural effects on relevant state and non-state actors'; and impact, being 'the effects on the ultimate target of governance'.[39] The indicators to determine the existence of synergies, settled on by van Asselt, are shared principles and concepts (to the extent that such are observable); common economic incentives (promoting the same types of activities); streamlined monitoring and reporting obligations; shared supporting measures; and 'positive' diffusion and learning.[40] The application of these indicators is considered in the next section.

THE MARKET MECHANISM COMPARTMENTALIZED

Before attempting to apply van Asselt's indicators, it is helpful to briefly recap. First, from the foregoing it can be seen that there is an issue of fragmentation in international law. Second, the consequences of fragmentation might be considered in terms of the management of interrelationships between regimes, where regimes are sets of norms, decision-making procedures, and organizations coalescing around functional issue-areas and dominated by particular modes of behaviour, assumption, and biases. While no classification for these interactions has been identified, this book argues that any such classification

---

[35] Ibid, 54.
[36] Ibid.
[37] Ibid.
[38] Ibid, 55–6.
[39] Ibid, citing fn.25 (Oberthür and Gehring) 34.
[40] Ibid, 56–7.

might also include the absence of interaction (thus, going beyond synergistic, conflicting, or neutral) as a consequence of fragmentation, as a category, especially where that interaction would otherwise further the objectives of that regime. It is noted that fragmentation can have consequences for both hard and soft law regime interactions. The object of the interaction – or absence thereof in this case – is the IET regime.

Third, it can be seen that international environmental law is fragmented, such being said to be epitomized by the international legal regime for climate change.[41] Academic writers have described this variously in terms of its complexity, and the range and number of measures, initiatives, and organizations that are now part of this regime. For instance, van Asselt and Zelli[42] have described this fragmentation in terms of a transition from the initial centrality of global climate governance under the UNFCCC, to a plurality of global climate governance involving not just the UNFCCC, but a multitude of non-UNFCCC governance arrangements, including international organizations such as the World Bank and other environmental treaty organizations; high-level, club-like forums such as the Group of 8 highly industrialized countries (G8), the Group of 20 major economies (G20), and initiatives of consecutive US Presidents Bush and Obama; various government/non-government stakeholder partnerships such as the Carbon Sequestration Leadership Forum and the Renewable Energy and Energy Efficiency Partnership; regulatory and voluntary emissions trading markets; corporate self-regulatory schemes such as Carbon Disclosure Project; and the various sub-national initiatives at provincial, city, and local levels of government.[43]

Keohane and Victor consider that what governments have created is '… a varied array of narrowly-focused regulatory regimes …' that they call the regime complex for climate change,[44] the components of which resemble the components of van Asselt and Zelli's plurality of global climate governance. Carlarne, on the other hand, looks at the climate issue from the perspective of its complexity and the inability of international environmental law to provide a satisfactory framework within which that complexity can be addressed.[45] In

---

[41] Cinnamon Carlarne, 'International Treaty Fragmentation and Climate Change' in D. A. Farber and M. Peeters (eds.), *Encyclopedia of Environmental Law: Volume 1 Climate Change Law* (Edward Elgar, 2016) 265.

[42] Harro van Asselt and Fariborz Zelli, 'Connect the Dots: Managing the Fragmentation of Global Climate Governance' (2014) 16(2) *Environmental Economics and Policy Studies* 137, 139–43.

[43] Also Robert O. Keohane and David G. Victor, 'The Regime Complex for Climate Change' (2011) 9(1) *Perspectives on Politics* 7, 10–12.

[44] Ibid, 7.

[45] Cinnamon Carlarne, 'Delinking International Environmental Law & Climate Change' (2014) 4 *Michigan Journal of Environmental & Administrative Law* 1.

Carlarne's view, the framing of climate change as an environmental law issue is flawed.[46] 'Climate change is a problem firmly rooted in our basic post-war, global economic model, a model that is based on an underlying assumption that free trade and economic growth can simultaneously improve global economic welfare and address distributive justice concerns'.[47]

This book takes another perspective, examining just a single element of Carlarne's complexity, of van Asselt and Zelli's plurality, or of Keohane and Victor's regime complex, namely IET, as provided for by Article 17 of the Kyoto Protocol. This single mechanism is considered from the perspective of the consequences of being compartmentalized inside the climate regime. As such, the consideration here of fragmentation is not looking at treaty conflict in the sense examined by the ILC, nor perhaps even looking at hard law. It is looking at a mechanism inserted in the Kyoto Protocol in order to improve the economic efficiency and cost effectiveness with which the climate regime might achieve its objective of mitigating greenhouse gas (GHG) emissions, and ultimately, the objective set out in Article 2 of the UNFCCC.[48]

The argument propounded is that IET being compartmentalized in the climate regime has had a negative consequence, namely that any potentially beneficial outcomes that might have been realized through synergies between IET and other financial markets in general, have not been so realized. As a consequence, IET under the Kyoto Protocol has been less effective furthering the UNFCCC objectives.

This argument looks not at the effect of fragmentation on international law generally, or on the climate regime as a whole, but rather having accepted that the climate regime is fragmented and made up of a multitude of components, by considering the impact of compartmentalization on a single element in that fragmented climate regime. Second, it considers not how fragmentation of the climate regime has led to either conflicting or synergistic interactions between different fragmented components, but rather by considering how compartmentalization of that element of the climate regime has limited interaction, preventing beneficial synergies which otherwise may have developed and enabled that component of the climate regime to better achieve the objectives for which it was put in place.

Providing evidence to support this argument is a difficult proposition, as flagged earlier by van Asselt. It requires first, analysis of the actual effectiveness of IET under the Kyoto Protocol; second, speculation as to the synergies

---

[46]   Ibid, 4.
[47]   Ibid, 13.
[48]   United Nations Framework Convention on Climate Change, 9 May 1992, 1771 UNTS 107.

that might have been achieved with financial markets more generally, had IET not been cocooned inside the climate regime; and third, what the aggregate effectiveness might have been had those synergies, in fact, occurred. Nevertheless, it is proposed to apply the framework elaborated by van Asselt, in an attempt to demonstrate this proposition. To reiterate, for this exercise, IET is treated as principally consisting of trade in Assigned Amount Units (AAUs) and Certified Emission Reductions (CERs), and trading under the European Union Emission Trading Scheme (EUETS). The ambit of the expression financial markets is elaborated in subsection 2 below.

1.   Effectiveness of IET under the Kyoto Protocol

As noted earlier, IET relies on the fundamental tenet of the exploitation of cost heterogeneity to minimize overall compliance costs.[49] Also noted earlier, mitigation of GHG emissions means changing behaviours, changing the way many economic activities are carried out and so, consequently, reducing their impacts.[50] Imposing a climate change related price on emissions is one way to do this, by way of emissions trading schemes (ETSs). The larger the market, the greater the efficiency benefits it might be expected to yield.[51]

Thus, it was envisaged IET would establish a carbon price that would determine the point at which emitters of GHGs would find it more cost effective to alter behaviour and reduce emissions. For a market to function there needs to be demand and supply. As no natural demand exists for the Kyoto Protocol units to be traded under IET, principally AAUs and CERs,[52] demand could only be driven by compliance. While a compliance mechanism was established under the Kyoto Protocol,[53] it was unrealistic to expect that this would drive demand in a market that consisted of the less than 40 developed country parties (and EU), listed in Annex B of the Protocol (Annex B Parties) as having

---

[49]   Shi-Ling Hsu, 'International Market Mechanisms' in C. Carlarne, K. Gray, and R. Tarasofsky (eds.), *Oxford Handbook of International Climate Change Law* (Oxford University Press, 2016) 241 citing William J. Baumol and Wallace E. Oates, *The Theory of Environmental Policy* (2nd edn, Cambridge University Press, 1988) 21–3; and Tom Tietenberg and Lynne Lewis, *Environmental and Natural Resource Economics* (10th edn, Pearson, 2014) 357.

[50]   Justin D. Macinante, 'A Conceptual Model for Networking of Carbon Markets on Distributed Ledger Technology Architecture' [2017] *CCLR* 243.

[51]   Fn.49 (Hsu).

[52]   Emission Reduction Units (ERUs) and Removal Units (RMUs) were also tradable, but trade in these was very small in comparison to the other units, especially CERs.

[53]   Decision 27/CMP.1, FCCC/KP/CMP/2005/8/Add.3; the objective of the procedures and mechanisms under which the Compliance Committee is established are: 'to facilitate, promote and enforce compliance with the commitments under the Protocol'.

quantified emission limitation reduction commitments (QELRCs) that ratified it, especially when they had the option to leave the process if they so desired (which Canada ultimately chose to do). The only realistic way for IET to work, it is submitted, would be for those Annex B Parties to establish domestic markets, but with the notable exception of the EU, this did not occur in any significant way. The World Bank reflected, in 2014, that: 'It cannot be assumed that sovereign players will use a marketplace. The trading activity is primarily driven by private sector players. If a market, such as the AAU market, is to be made effective there may need to be an explicit role for the private sector'.[54]

The AAU market was for countries with QELRCs 'to achieve these at least cost'.[55] However, from an early stage it was observed 'that even with trading, the heterogeneity of national policies would mean that AAU trading could not achieve a least-cost outcome', as countries would make policy decisions based on national priorities and 'would not necessarily optimise on carbon price alone'.[56] Furthermore, economic collapse in the former Eastern bloc countries (the economies-in-transition (EITs)), in the late 1990s, provided an immediate supply of AAUs not stemming from any specific mitigation activity.[57] The absence of transparent procedures in relation to purchases of Ukrainian AAUs introduced reputational risk.[58] Most of the AAU purchasers, in the end, were from the Japanese private sector, sovereign purchasers from Spain and Austria, and the World Bank.[59]

The market in CERs has been more successful in terms of the levels of activity it has generated. The World Bank indicates that 1,155 billion CERs had been issued by the end of 2012 (that is, over about eight years) and that the Clean Development Mechanism (CDM) had generated US$130 billion of investment in GHG reducing project activities.[60] Yet, as noted earlier, the CDM has been beset by numerous design issues and has been in a constantly evolving state.[61]

---

[54] World Bank, Ecofys, *State and Trends of Carbon Pricing 2014* (Washington, DC, 2014) 44. Also the EBRD observes that one of the challenges for Article 6, Paris Agreement to succeed is restoring private sector confidence in policy dependent markets, see: European Bank for Reconstruction and Development (EBRD), Operationalising Article 6 of the Paris Agreement: Perspectives of developers and investors on scaling-up private sector investment, May 2017 <www.ebrd.com> accessed 21/09/17.

[55] Ibid (World Bank 2014).
[56] Ibid.
[57] Ibid.
[58] Ibid.
[59] Ibid.
[60] Ibid, 44–5.
[61] Ibid.

The EUETS has dominated the carbon market and continues to do so. For example, in 2008 it recorded transactions valued at US$92 billion, a year-on-year growth of 87 per cent from 2007. However, the effects of the economic slow-down were beginning to be felt already at that time and as demand and commodity prices collapsed, so did emissions.[62] Companies already holding excess European Union Allowances (EUAs) due to previous over-allocation, sold in order to raise cheap (essentially free, since allocations had been free) cash in an illiquid environment, and consequently prices fell dramatically.

While noting the growth globally in implementation of carbon pricing mechanisms in various jurisdictions, both national and sub-national, the World Bank reported in 2014 on setbacks for IET under the Kyoto Protocol, with Canada having withdrawn during the first commitment period, the Australian government planning to repeal its carbon pricing mechanism, and Japan, Russia, and New Zealand officially pulling out of the second commitment period.[63] Market infrastructure was continuing to be dismantled as many market participants including banks, private sector intermediaries, and aggregators, and Designated Operational Entities (DOEs) under the CDM, had either already exited or were substantially reducing their activities and exposure to the market.[64] It was reported that the absence of any clear policy signals had led to fears that 'demobilisation of the CDM market infrastructure' would 'substantially damage the institutional memory' that had been created.[65] Additionally, confidence in the EUETS had been negatively impacted by the mechanism's design inability to deal with the market downturn. Consequently, prices for EUAs dropped substantially and CERs became almost worthless.[66]

(i)   Output

Turning now to apply the three levels, flagged by van Asselt, by which a regime's or, in this case, a mechanism's effectiveness might be measured, the first is that of output, considered in terms of the norms generated, prompting the question what norms, or accepted standards of behaviour, have been generated by IET?

The information on the status of IET presents a mixed picture. While the level of carbon price might also be considered as an outcome of IET, if the purpose of IET under Article 17 of the Kyoto Protocol was to establish

---

[62]   World Bank, *State and Trends of the Carbon Market 2009* (Washington, DC, 2009) 5–7.
[63]   Fn.54 (World Bank, 2014) 16.
[64]   Ibid.
[65]   Ibid.
[66]   Ibid, 17.

a carbon price, which would determine the point at which emitters of GHGs would find it more cost effective to alter behaviour and reduce emissions, then it seems, on balance, to have been ineffective. The AAU market has proved to be ineffective. As noted, AAU trading could not achieve a least-cost outcome. The CER market has generated a much greater level of activity, but while it could be said to be generating norms for projects that produce credits, it suffers due to design issues, frequent rule changes, policy changes impacting the usability of CERs, and market uncertainty over its future. Based on the information cited, the CER market has been ineffective in establishing a carbon price, as the CER price seems to be determined largely by the EUA price. The EUETS market provides the major component of IET globally and while it showed much promise in the years up to the time when the impact of the global financial crisis was manifest, design flaws and operational issues have undermined market confidence in it. Nevertheless, it might be seen as effective, to a degree, in establishing a carbon price and thus helping to establish a norm for factoring carbon emission cost into business decision-making (at least in the EU).

(ii)   Outcome

The second measure of effectiveness flagged is outcome, described as being behavioural effects on relevant state and non-state actors, corresponding to behavioural interaction. If this measure is assessed in terms of the growth in implementation of ETSs and other carbon pricing measures in jurisdictions, both national and sub-national, around the world, then it might be concluded that IET has been effective. About 100 of the signatories to the Paris Agreement stated in Intended Nationally Determined Contributions (INDCs) that they are considering or planning to put a price on carbon,[67] although not all of these involved implementing market measures.

Over 1200 companies worldwide indicated that they are using or plan to use internal carbon pricing in the next two years and, of these, 83 per cent are located in jurisdictions with (scheduled) mandatory carbon pricing initiatives,[68] although it is not immediately apparent how many of these jurisdictional carbon pricing initiatives involve a market mechanism. All the same, numerous public-private initiatives have been implemented,[69] including the

---

[67]   World Bank, Ecofys and Vivid Economics, *State and Trends of Carbon Pricing 2016* (Washington, DC, 2016) 22 <https://openknowledge.worldbank.org/bitstream/handle/10986/25160/9781464810015.pdf?sequence=7&isAllowed=y>   accessed 28/06/17. These INDCs cover about 58 per cent of global GHG emissions.

[68]   Ibid, 24.

[69]   Ibid, 30.

Carbon Pricing Leadership Coalition,[70] the Carbon Disclosure Project[71] and The Carbon Pricing Panel,[72] the G7 Carbon Market Platform,[73] and the High Level Panel on Carbon Pricing.[74]

In these terms, it would seem that there have been significant behavioural effects on relevant state and non-state actors, although it is hard to pin down precisely the extent to which this can be ascribed to IET under the Kyoto Protocol. Nevertheless, to a certain degree it would seem to be an extension of what started there. On the other hand, if the behavioural effect in terms of which the outcome of IET is to be assessed is a change in emissions behaviour then, just as for the third measure (following), it would seem to have been ineffective. Also, as noted earlier, carbon price might be considered to be an outcome, rather than an output, while similarly, the various regulatory structures that have developed under the Kyoto Protocol are seen as outputs. Notwithstanding such alternative approaches, the result in terms of an assessment of the mechanism's effectiveness would be the same.

(iii)   Impact

The third measure is impact, being the effects on the ultimate target of governance. Here, if the target of governance of IET was to achieve lower costs emission reductions, then (assuming the economic theory is correct) it can be argued that IET has had an impact. However, if on the other hand the ultimate target of IET is accepted as having been to mitigate global GHG emissions – that is, not just to do so at lower cost but actually to achieve the emissions reductions, at lower cost – then, in this respect, one might conclude it has been largely ineffective. The evidence indicates that, rather than the rate of emissions decreasing, it has increased since IET was implemented. For example, it has been reported that about half the anthropogenic carbon dioxide (CO_2) emissions between 1750 and 2011 occurred in the last 40 years.[75] Total GHG emissions increased on average by 1.3 per cent per annum over the period 1970–2000; for the period 2000–10, however, the average annual increase jumped to 2.2 per cent, despite a growing number of climate change mitigation

---

[70]   See <https://www.carbonpricingleadership.org/> accessed 21/09/17.

[71]   See <https://www.cdp.net/en/> accessed 21/09/17.

[72]   See <http://www.worldbank.org/en/news/speech/2016/04/21/carbon-pricing-panel---setting-a-transformational-vision-for-2020-and-beyond> accessed 21/09/17.

[73]   See <http://www.bmub.bund.de/en/topics/climate-energy/climate/international-climate-policy/carbon-market-platform/> accessed 21/09/17.

[74]   See <https://www.carbonpricingleadership.org/report-of-the-highlevel-commission-on-carbon-prices/> accessed 21/09/17.

[75]   Intergovernmental Panel on Climate Change (IPCC), Fifth Assessment Report, Climate Change 2014 Synthesis Report, Summary for Policymakers <http://www.ipcc.ch/pdf/assessment-report/ar5/syr/AR5_SYR_FINAL_SPM.pdf> accessed 20/04/17, 4.

policies.[76] The Intergovernmental Panel on Climate Change (IPCC) noted: 'Globally, economic and population growth continued to be the most important drivers of increases in CO2 emissions from fossil fuel combustion. The contribution of population growth between 2000 and 2010 remained roughly identical to the previous three decades, while the contribution of economic growth has risen sharply'.[77]

It is noted that the 2000–10 period includes at least two years during which the global economy laboured under the impact of the global financial crisis. Whether, in these circumstances, IET had any impact on emissions is impossible to say definitively, and not even approximately without detailed mathematical and statistical analysis of the data, which is beyond the scope of this book. However, based on the simple fact of the rate of increase in emissions for part of the period during which IET has operated, one might conclude that it is difficult to ascribe any impact to IET in terms of effectiveness, on this basis, covering both outcome and impact.

2.   Synergies That Might Have Been Achieved

The second step in this process speculates on what synergies might have been achieved with financial markets more generally had IET not been cocooned inside the climate regime. The following quote from a time when the outlook for the carbon market was positive, before the impact of the global financial crisis was felt, is most pertinent:

> The world has truly changed today when power company executives and investment bankers talk about climate risk and environmentalists talk about leveraging the power of markets. Climate policy has mobilised the world of private capital to work in favour of protecting the environment. In doing so, it has brought together two widely different worlds with very little knowledge and experience of each other. A good example of the disconnect between the two worlds was the unauthorised release of verified EU ETS emissions data in April 2006, which highlighted the need for environmental officials to safeguard emissions data, which, for the first time, had large financial implications.
>
> Each of the two worlds described above has very different mental models and very little knowledge about how the other world operates, let alone any deep insights into the other's assumptions, motivations, language, and behaviour. Considering how widely different these two cultures are, it is quite extraordinary to recognize how successfully they have worked together so far to produce concrete action to reduce carbon emissions. In 2007, some prominent investment banks tried to further bridge the gap between the two worlds, as they hired specialist carbon staff, bought small

---

[76]   Ibid, 5.
[77]   Ibid.

and boutique carbon originators and made investments in the 'infrastructure' of the carbon market, including exchanges and registries.[78]

While this carbon market utopia may have germinated in the private sector,[79] the same cannot be said for the public sector – the regulatory and institutional frameworks that were, after all, the designers, the managerial overseers, and drivers of market demand. The unauthorized release of data (the 'disconnect' cited in the quote) disclosed the fact, suspected by some already, that national emissions inventories had been over-estimated by national authorities in the EU, resulting in over-allocation of EUAs and consequently an over-supply in the market. The obvious outcome was a price plunge. As the World Bank observed, there were (and still are) two sides to the IET market, the environmental and the financial, and the 'very different mental models' referred to by the Bank did not stop with the market participants. The different mental models, lack of knowledge and insight applies also to the policymakers (negotiators), regulators, and administrators of the IET market.[80]

To consider what synergies might have been achieved had IET not been cocooned in the climate regime, it is proposed, so far as is possible, to apply the indicators for determining the existence of synergies from van Asselt (see preceding section): shared principles and concepts (to the extent that such are observable); common economic incentives (promoting the same types of activities); streamlined monitoring and reporting obligations; shared supporting measures; and positive diffusion and learning.

First, though, it is necessary to clarify what is meant here by the expression financial markets. As applied, this expression is intended to refer broadly to financial markets that trade internationally, such as the foreign exchange market, debt (bond) markets and some commodities markets. These are the sorts of markets with which, it is posited, the IET market might have achieved synergies, in a functional sense. As markets, the IET market and the financial markets share principles and concepts common to all markets: supply and demand, liquidity and depth, access to market information, and so on. How

---

[78] World Bank, *State and Trends of the Carbon Market 2008* (Washington, DC, 2008) 22.

[79] Flowering briefly, before shortly later beginning to wither. Note also that the issue of coordination between carbon market and financial market regulators was raised by the European Court of Auditors in its 2015 report on the EUETS.

[80] Another perspective on this issue might be in terms of engagement between the private sector and the UNFCCC, as examined in a report for the European Commission in 2010: World Business Council for Sustainable Development Secretariat, Ecofys and Climate Focus, 'Private Sector and the UNFCCC Options for Institutional Engagement', Final Report 31/8/2010 <https://www.wbcsd.org/Clusters/Climate-Energy/Resources/Options-for-institutional-engagement-in-the-UNFCCC-process> accessed 11/04/18.

could these shared principles and concepts have been better considered, had the IET market been exposed more to financial regulatory and institutional oversight, rather than just environmental? An obvious initial answer is in better understanding the importance of financially sensitive market information from the outset: the failure to do so has been noted in the quote above.

A second, related way would be in better appreciating the importance of balancing supply and demand, and hence better controlling the way in which national emissions data (which has a direct relationship both to the size of allocations and hence supply, and to the anticipated size of demand) was both compiled and managed overall throughout the process. A third way, related to the initial design, would have been in better understanding the need for liquidity and depth in the market, especially in achieving smoother price movements. If the EU had not implemented its ETS, thereby creating a market of over 11,000 potential traders, the market might only have consisted of the 37 Annex B Parties and any legal entities they authorized, trading mostly AAUs and CERs.

Synergies exist in terms of the other indicators as well. Both IET and financial markets have common economic incentives in the profit motive for participants, although in the case of IET, this is, or at least should be, subordinated to the overriding environmental objective of mitigating emissions, the reason for which the IET market exists. The IET primary markets (in AAUs, CERs, EUAs) have generally been seen less as financial markets, the units traded being considered more often as commodities rather than as financial instruments (although in the European Union, at least, this is changing).[81] However, just as there are for other markets, there have been derivatives markets in the units traded in the IET market and these have always been regulated as financial markets.

Hence, it seems opportunities exist for synergies in terms of streamlined monitoring and reporting obligations, shared supporting measures, and 'positive' diffusion and learning, were IET markets and financial markets more aligned. In particular, for instance, the closer oversight of the primary market in AAUs, CERs, EUAs and secondary (derivative) markets – futures, forward contracts, options – based on the underlying contracts for those units might have been expected to afford regulators (both environmental and financial) more of the '... deep insights into the other's assumptions, motivations, lan-

---

[81]   MiFID II defines 'financial instrument' to include emission allowances, consisting of any units recognised for compliance with requirements of Directive 2003/87/EC (Emissions Trading Scheme): Directive 2014/65/EU of the European Parliament and of the Council of 15 May 2014 on markets in financial instruments and amending Directive 2002/92/EC and Directive 2011/61/EU, OJ L 173, 12.06.2014, 394–496, Annex I, Section C, paragraph (xi), with effect from January 2018.

guage and behaviour' referred to by the World Bank. The EUETS is moving in this direction.[82]

Nevertheless, the damage appears to have been done in terms of the lost opportunity for greater synergies and in terms of private sector sentiment towards an international carbon market becoming more negative. Overall, it is concluded that IET has been, on balance, less effective than it could have been. Had it not been compartmentalized, there are synergies with other financial markets (and, probably, between primary and secondary carbon markets) that may have been realized. All the same, it is acknowledged these conclusions are, at best, speculative.

3.   Aggregate Effectiveness Had Those Synergies Occurred

To complete this exercise, it is necessary to consider what aggregate effectiveness might have been achieved, had those synergies, in fact, occurred. Again, it is acknowledged that this is speculative. Nevertheless, had the IET market been less compartmentalized inside the climate regime, it might be expected that there would have been more private financial sector involvement from the start, not just as market participants, but in shaping market design. If this had been the case, then it is reasonable to expect that there would have been a greater roll out of trading and, consequently a deeper and more liquid market, with a less volatile carbon price being more widely factored into business planning from the outset. To some degree, this private sector engagement was engendered by the early implementation of the EUETS, however, as also mentioned earlier, this too had problems.

Climate change needs to be made mainstream and '... released from a compartmentalised framing'.[83] One way proposed to achieve this is by framing climate change as an energy challenge.[84] However, this book argues that mainstreaming the climate change issue into national and international economic decision-making can be achieved more readily if international emissions trading is perceived and treated more as an international financial market.[85] The inclusion of the flexible mechanisms and especially IET, in the Kyoto Protocol, has already opened the door, at least partially, for this to happen.

---

[82]   In this respect, see the ECA observations re alignment of environmental and financial regulators: European Court of Auditors, 'The integrity and implementation of the EU ETS' Special Report, 2015, paragraphs 12–24 <http://www.eca.europa.eu/Lists/ECADocuments/SR15_06/SR15_06_EN.pdf> accessed 23/06/17.

[83]   Fn.45 (Carlarne) 48.

[84]   Ibid.

[85]   What this could entail has been elaborated in Chapter 1.

Aggregate effectiveness that might have been achieved, had the synergies between IET and the financial markets been realized, includes the possibility, for instance, that bond (debt) markets may have started factoring in a carbon price as part of the pricing of issuances. One could imagine this being the case for countries particularly exposed to fossil fuel resource exports (e.g., Australia) and for resources companies operating in those countries and borrowing in global financial markets. A second potential synergy might have been the carbon price being taken into account in commodities markets – not as a trade issue in the form of, say, a border tax adjustment – but upstream, being integrated into the cost of the commodity from the point of view of production, energy use, and transportation. Third, the carbon price might even have become another of the many considerations that currency traders take into account (perhaps only to a small degree, but considered nonetheless) when setting and adjusting currency exchange rates.[86]

4.   Conclusion on Compartmentalization

The argument made here is that the way in which IET evolved resulted in its being compartmentalized in the climate regime, and this has had a negative consequence. Potentially beneficial outcomes that might have been realized through synergies between IET (had it been treated as an international financial market) and other financial markets generally have not been realized. There has been a substantial loss of private sector confidence and support, and IET's effectiveness as a mitigation policy measure has been impaired as a consequence. The proposition is that, on balance, IET has been less effective than it could have been in furthering the UNFCCC objectives. Compartmentalization may have played a part in this outcome by impeding potential synergies that may have arisen between IET and financial markets.

It is acknowledged that the exercise carried out here does not apply scientific or statistical rigour. The conclusion, involving as it does a certain amount of speculation as to the outcomes that might have been achieved, cannot be any more definitive. All the same, this exercise serves to underscore the theme that international emissions trading can be more effective in achieving mitigation outcomes. The market – if well-designed – should operate as a global financial market, but should do so within an equally well-designed boundary framework of climate change rules that give effect to the intention of the parties to the

---

[86]   This might certainly be the case if a universally accepted methodology for determining mitigation value (MV) of different jurisdictions' mitigation outcomes could be settled upon: this is discussed later in this book in the context of the conceptual model proposed.

Paris Agreement. The market model proposed by this book aims, amongst other things, to address this and so to promote re-engagement of the private sector in the carbon market.

4. The nature of what is traded in the carbon market

As mentioned in the preceding chapter, the carbon market embodies dual functions, operating as both a regime to implement the environmental policy measure of mitigating greenhouse gas (GHG) emissions, and as a trading market. In relation to this duality, the preceding chapter considered the interaction of these functions from a broad perspective, arguing that compartmentalizing the carbon market in the climate regime has been detrimental to potentially beneficial outcomes that otherwise might have been realized. As a consequence, emission trading has been less effective as an environmental policy measure for mitigating GHG emissions than might otherwise have been the case.

This functional duality is pertinent also to the question of the nature of what is traded in the carbon market,[1] examined in this chapter, which focuses at a more granular level. It is argued that the way in which emissions trading evolved, particularly under the Kyoto Protocol, has resulted in greater emphasis and attention being placed on the entitlement, the nature of the rights associated with what is traded, rather than on its public policy and environmental function. This chapter aims to reconnect with that environmental function by arguing for a refocusing on to the value, rather than just the nature, of what is traded. It proceeds by reviewing briefly the background on the nature of what is traded in emissions trading schemes; it considers instances that have exposed the shortcomings in the approach to the nature of what is traded; then it analyzes the transition in the current international legal position from the homogeneous approach under the Kyoto Protocol, to the heterogeneity of mitigation outcomes recognized by the Paris Agreement.

The focus on the rights associated with what is traded has been framed in terms of a balance between the security afforded private law rights (in the

---

[1]   This corresponds to the theme of the need for greater certainty over the nature of allowances in the European Emission Trading Scheme, see European Court of Auditors, 'The integrity and implementation of the EU ETS' Special Report, 2015, paragraphs 12–24 <http://www.eca.europa.eu/Lists/ECADocuments/SR15_06/SR15_06_EN.pdf> accessed 23/06/17. The discussion in this chapter mostly focuses on allowances issued under emission trading schemes, although it is recognized that project-based credits are traded also.

emissions entitlements) and the need for regulatory flexibility (for example, in being able to adjust the caps).[2] This chapter argues that the advent of the Paris Agreement has changed the situation, not by altering the balance away from market certainty in favour of regulatory flexibility, or vice versa, but rather through reaffirmation of the purpose for which the trading market exists in the first place, namely, to achieve reductions of GHG emissions effectively and efficiently. Thus, it argues that it is not just the nature, but the value, in mitigation terms, of what is traded that needs to be the focal point in the new regime under the Paris Agreement.

APPROACHES DEVELOPED TO THE ISSUE

1.   The Focus on Characterizing What is Traded

The focus on the nature of what is being traded in emission trading schemes (ETSs) is understandable, given that its importance goes beyond just the operation of the schemes themselves. The Financial Markets Law Committee (FMLC) of the Bank of England has opined, in relation to the European Union Emissions Trading Scheme (EUETS) but with relevance to emissions markets generally, that without clarification of the legal classification of emission allowances, the issues identified by the FMLC '… could significantly impede upon the development of the market …'[3] There have been other, more existential impacts on the carbon market since that time, such as from the global financial crisis. All the same, the legal classification of emission allowances and rights attaching thereto remains an outstanding issue that may yet detract from the much needed, renewed basis for support, said to be afforded by the Paris Agreement.[4] The FMLC stated that the most significant ramification was the relevance of the legal nature of an emission allowance '… in determining which law properly governs the creation, transfer and cancellation of the allowance, and whether (and how) security rights can be created over that allowance'.[5]

---

[2]   Sabina Manea, 'Defining Emissions Entitlements in the Constitution of the EU Emissions Trading System' (2012) 1(2) *Transnational Environmental Law* 303, 308.

[3]   Bank of England, Financial Markets Law Committee, 'Issue 116 – Emission Allowances: Creating Legal Certainty', (October 2009), 1.6 <http://web.archive.org/web/20170108031056/http://www.fmlc.org/uploads/2/6/5/8/26584807/116e.pdf> accessed 11/05/17.

[4]   International Institute for Sustainable Development, Earth Negotiations Bulletin, Vol.12, No.663, Summary of the Paris Climate Change Conference, 29 November–13 December 2015, 45 <http://enb.iisd.org/download/pdf/enb12663e.pdf> accessed 26/06/17.

[5]   Fn.3 (FMLC) 1.5.

   ETSs are not ends in themselves, but are intended to generate a price signal, creating conditions whereby entities can plan long-term action to reduce or avoid emissions,[6] thus '... the question of whether carbon units are property concerns the important question of investment certainty'.[7] Hedges also notes that whether the traded unit is a form of property is deeply relevant to decisions in a system designed to drive business investment decisions.[8] Practical questions include, not only as to whether the asset can be used as security, but also whether the value is predictable, or whether the asset could be subject to unrestricted regulatory changes that affect its value.[9] Further issues include how allowances should be treated for tax purposes and for accounting purposes; how they should be dealt with in the case where the registered holder becomes insolvent;[10] how financial investment regulation might apply to allowances themselves, or derivative contracts in the allowances; and whether allowances could be subject to property-based criminal acts, such as theft.[11]

   Other authors also have flagged the general failure by ETSs to specify the nature or scope of the emission allowances they trade. Some point to lessons to be learnt from the global financial crisis of 2007–8, where the nature of what was being traded – in that case, securitized sub-prime mortgages – was inadequately regulated and poorly understood.[12] The nature and treatment of traded units is important for giving legal security and certainty to governments, private and public sector entities, and providing confidence to the trading system.[13] Concerns over the absence of an adequate definition addressing legal

---

   [6]   Andrew Hedges, 'Carbon Units as Property: Guidance from Analogous Common Law Cases' [2016] *CCLR* 190, 191.
   [7]   Ibid.
   [8]   Ibid.
   [9]   Ibid.
   [10]   Although not touched on by the FMLC, this might also depend on the nature of the registered holder: for instance, if the allowances are held for surrender against emissions for compliance purposes, would this make a difference from if they were held for investment purposes?
   [11]   Fn.3 (FMLC) 1.5, 2.7.
   [12]   Hope Johnson et al., 'Towards an International Emissions Trading Scheme: Legal Specification of Tradeable Emissions Entitlements' (2017) 34(1) *Environment and Planning Law Journal* 3. Although the securitized sub-prime mortgages would probably be a lot more complicated to decipher than emission allowances.
   [13]   Matthieu Wemaere, Charlotte Streck, and Thiago Chagas, 'Legal Ownership and Nature of Kyoto Units and EU Allowances' in David Freestone and Charlotte Streck (eds.), *Legal Aspects of Carbon Trading: Kyoto, Copenhagen and Beyond* (Oxford University Press, 2009) 36; this confidence and certainty also underpins development of any secondary (derivatives) markets: Andrew Hedges, 'The Secondary Market for Emissions Trading: Balancing Market Design and Market Based Transactions' in David Freestone and Charlotte Streck (eds.), *Legal Aspects of Carbon Trading: Kyoto,*

status of allowances under the EUETS (EUAs) were raised by the European Court of Auditors, which noted that the emissions market needs sufficient liquidity to function well, and that the EUETS could improve, in this respect, if there were to be more certainty over an EU-wide definition of EUAs.[14]

Much of the attention has been focused on whether or not emissions allowances constitute property and, if so, what sort of rights attach. The meaning of property rights is seen as being '... central to the language of economics ...'[15] and a system of property rights as forming the basis for all market exchange.[16] Consideration of rights attaching to emissions allowances must take account of the interaction between the public policy aims of the scheme (such as those of the EUETS), and the private law entitlements created by it,[17] and may not easily fall into pre-existing categories such as property rights or personal rights, but may well form a category of their own, which needs to be accurately defined.[18]

It can be seen that the nature of what is being traded in ETSs can have implications for the individual parties to transactions, not just in terms of the ability to use it as security, or for tax purposes, or for its accounting treatment, but also in resolving disputes when, for example, criminality intervenes (for instance, in the EUETS when the traded units have been stolen and on-sold to innocent third parties). From a broader perspective, how the public policy need for flexibility to make adjustments to emissions rights is balanced with the private law entitlements attaching to those emissions rights can have implications for the operation of an emissions market itself, in terms of ability to meet the policy objectives for which it has been established. The impact of this balancing exercise on market operation could have implications for the overall suite of public policies within which that mechanism resides. For example, Manea notes that the EUETS forms part of a larger scheme, the EU Climate and Energy Package, and thus 'does not exist in a regulatory void' where so long as reductions are achieved it matters not how they are.[19] Rather, emission reductions are part of 'concerted efforts to move to a low-carbon economy',

---

Copenhagen and Beyond (Oxford University Press, 2009) 314; also on this point generally, see: Charlotte Streck and Moritz von Unger, 'Creating, Regulating and Allocating Rights to Offset and Pollute: Carbon Rights in Practice' [2016] *CCLR* 178.

[14] Fn.1 (ECA) paragraphs 25–8.

[15] Daniel Cole and Peter Grossman, 'The Meaning of Property Rights: Law versus Economics?' (2002) 78(3) *Land Economics* 317.

[16] Ibid.

[17] Fn.2 (Manea).

[18] Ibid, 312.

[19] Ibid, 314.

and so need also to promote wider environmental policy goals such as green investment and an energy efficient economy.[20]

2.   Definitions in Legislative Schemes

Before global policymakers began to consider greenhouse gas (GHG) emissions trading and climate change, regulators in the United States (US) addressed the nature of what was being traded under the acid deposition control, or acid rain, program under the 1990 amendments to the US federal Clean Air Act:

> An allowance allocated under this subchapter is a limited authorization to emit sulfur dioxide in accordance with the provisions of this subchapter. Such allowance does not constitute a property right. Nothing in this subchapter or in any other provision of law shall be construed to limit the authority of the United States to terminate or limit such authorization ... Allowances, once allocated to a person by the Administrator, may be received, held, and temporarily or permanently transferred in accordance with this subchapter and the regulations ...[21]

The rationale for exclusion of 'property rights' in this and similar formulations under US federal and state laws has been ascribed to need to accommodate the protection of property from interference without compensation right under the US Constitution.[22] All the same, the formulation preserves rights to hold and transfer, which imply a form of property interest, namely, as against third parties, but not against the government.[23]

The Kyoto Protocol provides for each Annex B Party's assigned amount to be divided into assigned amount units (AAUs), defined in the modalities, rules

---

[20]   Ibid.

[21]   Acid Rain Program, 42 U.S.C. United States Code, 2011 Edition, Title 42 - The Public Health and Welfare, Chapter 85 - Air Pollution Prevention and Control, Subchapter IV-A - Acid Deposition Control, Sec. 7651b - Sulfur dioxide allowance program for existing and new units, (f) Nature of Allowances <https://www.gpo.gov/fdsys/pkg/USCODE-2011-title42/html/USCODE-2011-title42-chap85-subchapIV-A-sec7651b.htm> accessed 06/07/17.

[22]   Fn.2 (Manea) 316: Fifth Amendment. On this point, see generally: fn.13 (Wemaere et al.) 52; fn.13 (Streck, von Unger) 183–4.

[23]   Ibid (Wemaere et al). In US jurisprudence property rights are based on a relational approach meaning that to establish a right (as opposed to a lesser interest), one must be able to identify a corresponding duty that another owes in relation to that right, thus a legally enforceable right presumes a legally enforceable duty: fn.15 (Cole and Grossman) citing Wesley N. Hohfeld, 'Some Fundamental Legal Conceptions As Applied In Judicial Reasoning' (1913) 23 *Yale L.J.* accessed 20/02/19.

and guidelines for emissions trading,[24] as being '... equal to one metric tonne of carbon dioxide equivalent, calculated using global warming potentials defined by decision 2/CP.3 or as subsequently revised in accordance with Article 5'.[25] The Kyoto Protocol units are, first, accounting units, but also represent an entitlement to release an equivalent amount of GHGs, which is transferrable.[26] It has been observed that, rather than providing clear definitions of what they are, schemes normally describe what allowances entitle the holder to do.[27]

The EU has been criticized in relation to the definition of allowances (EUAs) under the EUETS.[28] Initially, it had been proposed to define an EUA as an 'administrative authorization',[29] but with the decentralized approach taken this was rejected in favour of each EU member state making a determination under its own national law.[30] Despite EUAs being described as having legal characteristics that could be viewed as 'property rights', their status in this respect seems to remain ambiguous across the jurisdictions,[31] resulting in an array of legal classifications by individual member states:[32] for example, Wemaere et al. note that an EUA was deemed a financial instrument in Sweden, but treated as a tradable commodity in Austria, Germany, France, Italy, Poland, Portugal,

---

[24] Decision 11/CMP.1, FCCC/KP/CMP/2005/8/Add.2, 17 <http://unfccc.int/ resource/docs/2005/cmp1/eng/08a02.pdf#page=17> accessed 06/06/17.

[25] Ibid, paragraph 3; similar definitions are set out for the other tradable units, certified emission reductions (CERs) under the Clean Development Mechanism, emission reduction units (ERUs) under joint implementation and removal units (RMUs) from land use, land-use change and forestry activities (LULUCF).

[26] Fn.13 (Wemaere et al.) 37.

[27] Ibid, 44, citing Jillian Button, 'Carbon: Commodity or Currency? The Case for an International Carbon Market Based on the Currency Model' (2008) 32(2) *Harvard Environmental Law Review* 571, 574.

[28] Fn.1 (ECA).

[29] Markus Pohlmann, 'The European Union Emissions Trading Scheme' in David Freestone and Charlotte Streck (eds.), *Legal Aspects of Carbon Trading: Kyoto, Copenhagen and Beyond* (Oxford University Press, 2009) 350.

[30] Ibid: this was apparently due to the perceived conflict with the EU principle of subsidiarity. The general aim of the principle of subsidiarity is to guarantee a degree of independence for a lower authority in relation to a higher body or for a local authority in relation to central government. It therefore involves the sharing of powers between several levels of authority, a principle which forms the institutional basis for federal States: http://www.europarl.europa.eu/aboutparliament/en/displayFtu.html?ftuId=FTU_1.2.2.html, accessed 03/08/15. 'The EU has now moved to the centralised allocation model, with an EU-wide cap on greenhouse gas allowances': Marjan Peeters, 'Greenhouse Gas Emissions Trading in the EU' in D. A. Farber and M. Peeters (eds.), *Encyclopedia of Environmental Law: Volume 1 Climate Change Law* (Edward Elgar, 2016) 381.

[31] Fn.29 (Pohlmann) 351–2.

[32] Fn.13 (Wemaere et al.) 50.

and Spain; for accounting purposes, Spain, Finland, Italy, Malta, and Portugal treated EUAs as either intangible assets or financial instruments, while France, Netherlands, and Germany stipulated they be recorded as tangible assets or inventory.[33] The situation was summed up as being that the EUA '… does not fit easily in any legal system of the EU Members. It can be deemed as a right "sui generis" in many jurisdictions …'[34] It is noted that, as of January 2018, emission allowances (defined to include Kyoto project-based credits that are accepted for compliance purposes in the EUETS, as well as EUAs) are classified as financial instruments,[35] flagging a change to a more consistent approach.

Finally in this respect, a study of 23 ETSs in force or under consideration, including seven from China, found 21 defined the emissions units they traded by objective features, rather than by legal relationships they were able to support.[36] Thus, the authors concluded, holders of the units could not be certain of their ownership, nor of the rights associated therewith.[37]

3.   Some Considerations Arising from Common Law

In English courts at least, emission allowances would be likely to be treated as constituting property[38] based on decisions in relation to milk quotas, *Swift and Another v Dairywise Farms Ltd., and Others*,[39] and waste licenses, *Re Celtic Extraction Ltd (in liq); Re Bluestone Chemicals Ltd (in liq)*.[40] In the *Re Celtic* case, the Court of Appeal applied the test set out in *National Provincial Bank Ltd v Ainsworth*,[41] that before a right can be recognized as being proprietary, or as affecting property, it must be definable, identifiable by third parties, capable of being assumed by third parties, and as having a degree of permanence or stability.[42] The waste licenses in *Re Celtic* were found to be property for the

---

[33]   Ibid, 51.

[34]   Ibid, 52.

[35]   Directive 2014/65/EU of the European Parliament and of the Council of 15 May 2014 on markets in financial instruments and amending Directive 2002/92/EC and Directive 2011/61/EU, OJ L 173, 12.06.2014, 394-496, Annex I, Section C, paragraph (xi), with effect from January 2018. This is discussed further in later chapters.

[36]   Fn.12 (Johnson et al). The schemes reviewed in this study were from the International Carbon Action Partnership (ICAP) Emissions Trading Worldwide, Status Report 2016.

[37]   Ibid, at 11.

[38]   Fn.3 (FMLC) 3.4.

[39]   [2000] 1 W.L.R. 1177.

[40]   [2001] Ch 475.

[41]   [1965] AC 1175.

[42]   Ibid, 1247–8.

purposes of being disclaimed as onerous by a liquidator, within the meaning of insolvency legislation. The case has been described as one of the leading English law cases as it '... captured broad principles to be used in assessing whether a statutory instrument can be a form of property'.[43] The court arrived at three criteria for an administrative permit or statutory instrument to constitute a form of property, being first, existence of a statutory framework conferring an entitlement on a person or entity that satisfies certain conditions (even if the framework contains discretionary elements); second, the permit or instrument must be transferable; and thirdly, it must have value.[44]

The case of *Armstrong DLW GmbH v Winnington Networks Ltd*[45] is directly on point as it considered in detail the nature, as property, of EUAs under the EUETS. This case related to the theft from a registry account of EUAs and their on-sale to an unsuspecting purchaser. There was no dispute between the parties that EUAs were capable of constituting, and did constitute, property as a matter of law; what was in issue was their precise nature and characterization as property.[46] The judge reviewed the nature of property, citing the test in *National Provincial Bank Ltd v Ainsworth* and setting out the categories of property recognized by English law, before considering the precise nature of an EUA, which he found was not a right, in the sense that it does not give the holder a right that is enforceable by civil action.[47] Rather, it represented a permission,[48] or an exemption from a prohibition or fine. An EUA was property at common law, in terms of the Ainsworth test, and because it only existed electronically, was intangible property.[49] By applying the threefold test in *Re Celtic*, the conclusion was that an EUA was certainly 'property' and intangible property under the statutory definition in place.[50]

The *Armstrong v Winnington* case is one of a number reviewed by Hedges[51] in considering how common law principles have been applied to various different statutory instruments when considering whether they constitute a form of property and what rights attach thereto. Hedges' concern is that holders of carbon units (that is, emission allowances/entitlements) have the benefit of

---

[43] Fn.6 (Hedges).
[44] Fn.40 (*Re Celtic*) 488–9 (paragraph 33).
[45] [2012] EWHC 10 (Ch.).
[46] Ibid, paragraph 40.
[47] Ibid, paragraph 48.
[48] The judge here refers to it being a 'liberty in the Hohfeldian sense', that is, something less than a right, drawing a parallel with the US jurisprudential approach: see fn.23 supra; in relation to US cases, see the review of US Acid Rain Program cases by Manea: fn.2.
[49] Fn.45 (*Armstrong v Winnington*) paragraph 52.
[50] Ibid, paragraph 58.
[51] Fn.6 (Hedges).

legal protections applicable to property, as this would enhance investment and market liquidity. The cases reviewed are divided into, on one hand, those where the '… statutory instrument is a form of property sufficient to enliven the wider protections and restrictions at law' and on the other, where system design ensures power to adjust without transgressing compensable rights.[52] This parallels Manea's argument that interdependency between the viability of the emissions market and the successful pursuit of its environmental objective requires a careful balancing of, on one hand, the need for some level of security for emissions entitlement holders, and on the other, the need for regulatory flexibility to adjust the emissions cap as required for the purpose of the environmental policy objective.

Hedges and Manea are both concerned with the need for a clear definition of the emission entitlement and clarity as to regulatory interventions that might impact on the emission entitlement, whether it is legally considered to be intangible property (e.g., in English law) or a limited authorization not constituting a property right (e.g., US federal Clean Air Act). It is clear from their analysis and the cases considered that the focus is on what the holder of these entitlements has and is consequently permitted to do; and what the relevant governmental authority may do in relation to that holder and their entitlement. In this latter respect, Manea refers to allowance issuing authorities cancelling valid allowances when they deem necessary,[53] while Hedges refers to unfettered legislative power to adjust the regime without triggering claims for compensation or unlawfulness.[54]

It is understandable that legislatures, in creating emissions trading schemes, would want to reserve broad discretion to adjust, deal with, and amend the schemes to suit their policy objectives. All the same, it might be observed as perverse that a legislature would establish a scheme under which allowances or entitlements are distributed to entities (either for free or for payment) who must surrender them against their emissions, only then to forfeit or cancel the same before they can be surrendered as intended. To do so would seem to defeat the purpose of the scheme (suggesting poor design in the first place), since the value in the entitlements derives from their surrender under the scheme anyway.

Notwithstanding this, the issue of the legal nature of the entitlement and the rights attaching thereto is important. However, to consider an alternative approach, what if the potential or risk of forfeiture or cancellation of entitlements and related rights were to be removed entirely from the calculation?

---

[52]   Ibid, 199.
[53]   Fn.2 (Manea) 309.
[54]   Fn.6 (Hedges) 200.

A legislature might still make adjustments, as necessary, to fine tune its scheme in order to better achieve the environmental objectives, simply that those adjustments would not, or at least would not need to, include the possibility of forfeiture or cancellation. Such an approach would ameliorate, or even eliminate, the property and property rights issue in relation to the nature of the entitlements. How might such an alternative approach come about? It could operate, it is posited, by recognizing that the emissions entitlements have two values that can fluctuate: a financial value in the market, and an environmental value – not the static value determined by definition to always be, for example, one tonne avoided carbon dioxide equivalent GHG, but rather a value that would fluctuate according to the physical outcome ascribed to it (whether that be one tonne avoided GHG emission, or some other value) from time to time, based on periodic evaluations. Changes in this environmental value would then influence changes in the market price, so that the market would better reflect the physical outcome referable to the emissions entitlement.

A market based on such an approach to emissions entitlements can be compared with other financial trading markets, such as the bond (debt) markets or even share (equity) markets. These markets are not, it is conceded, the same as emissions markets (and, for example, the comparable regimes cited by Manea).[55] However, they do demonstrate how the market price fluctuates according to other values: for example, in the case of bonds, the price will depend on the likelihood of the issuer of the debt making timely payments to service that debt and repaying the capital at the end of the term of the bond. This, in turn, depends on a range of factors, both internal to the issuer (for instance, management, governance, cost controls) and external environmental factors (for instance, the markets in which it earns income) influencing its performance and ability in earning the revenue necessary to be able to service, then repay, the debt. The external factors can include (assuming the debt issuer is corporate, not governmental) actions of governments in the areas where the issuer operates. These factors, both internal and external, are constantly assessed by analysts, whose reactions to them result in the price movements seen in the debt market. In the case of shares, again the market price will fluctuate according to a range of factors, both internal to the company (again for instance, management, governance, cost controls) and external environmental factors (for example, the markets in which it earns income) influencing its ability to earn revenue to build the wealth of the company and also pay dividends. In light of their dual nature, emissions entitlements also correspond more to commodities or financial assets, than with administrative permissions or other statutory instruments, as the emissions market includes participants

---

[55]   Fn.2 (Manea).

that do not have a compliance-based relationship with the regulator, but rather engage in the market for investment or speculation.[56]

4.   The Issue of Hot Air and Determining the Value

Manea gets close to this issue of the value of emission entitlements in discussing the case of the Corus steelmaking plant in the UK.[57] In 2010, this plant had been mothballed and was to be sold to an investor, but was still set to receive a substantial number of EUAs under the EUETS. The situation gave rise to a number of issues, in particular, the fact that with the plant mothballed, Corus was effectively receiving an over-allocation of allowances that would be surplus to its needs. With little or no production, there would be virtually no emissions against which the allowances would need to be surrendered, meaning they could be included in the sale of the plant, or sold on the market, either way for a windfall profit. In the outcome, the UK government treated the allocation already issued for the year in question as being the property of Corus,[58] and future allocations were to be based on the extent of regulated activities (giving rise to emissions) carried on at the plant.[59]

This issue has arisen before: in the case of economies in transition (EITs) under the Kyoto Protocol as noted earlier,[60] their respective assigned amounts far exceeded their requirements for compliance over the first commitment period,[61] thereby providing them with immediate surpluses of assigned amount units (AAUs) that they could sell. The mothballed Corus plant case and the example of the EITs illustrate the limited and awkward options available to authorities under the Kyoto Protocol model addressing the issue of hot air allowances already issued. Should they risk legal action for cancelling future projected allocations? Since the Corus case, the European Commission has issued a decision on hot air allocations,[62] although as the current over-supply

---

[56]   Ibid, 321.

[57]   Ibid, 311–13.

[58]   Ibid: described by Manea as a 'debatable choice of words'.

[59]   Ibid.

[60]   See Chapter 2.

[61]   Pursuant to Article 3, Kyoto Protocol to United Nations Framework Convention on Climate Change, 11 December 1997, 2303UNTS162 (2005).

[62]   Commission Decision of 27 April 2011 determining transitional Union-wide rules for harmonized free allocation of emission allowances pursuant to Article 10a of Directive 2003/87/EC of the European Parliament and of the Council (notified under document C(2011) 2772), Articles 21 (significant capacity reduction), 22 (Cessation of operation of an installation), 23 (Partial cessation of operation of an installation): https://eurlex.europa.eu/LexUriServ/LexUriServ.do?uri=OJ:L:2011:130:0001:0045: EN:PDF, accessed 25/02/19.

situation in the EUETS and creation of a market stability reserve demonstrate, dealing with allocated EUAs, or projected allocations, when emissions are reduced due to economic factors, rather than abatement measures, remains problematic.

The hot air issue can be considered not just in terms of allowances as property and how action taken by authorities might affect the legal rights therein. The issue goes to the credibility of the trading regime and might be considered, alternatively, in terms of what is the appropriate value for the emission allowance unit, at any particular time, given the prevailing circumstances. Hence, rather than legislators and judges trying to deal with how to clarify the legal nature of allocated units, so as to balance certainty of private legal rights with flexibility in seeking public policy aims,[63] for units with a defined fixed value,[64] the alternative model envisaged proposes that the value of the traded emission allowance units fluctuate according to accurate, periodic assessment of those units' actual worth, in much the same way as the value of an asset in another financial market, such as the debt or equity markets might do.

Such a market model might provide greater flexibility in pursuing public policy aims, without transgressing unit holders' private legal rights, provided there was a clear process to set the value and information on that value setting transparently available to the market. This could be an avenue the Subsidiary Body for Scientific and Technological Advice (SBSTA) might investigate in developing guidance for operationalizing Article 6. As such, the next section looks at the current international legal context of the transition from the Kyoto Protocol to the Paris Agreement.

MOVING FROM A HOMOGENEOUS TO A HETEROGENEOUS APPROACH

1. Kyoto Protocol and Homogeneity

Under the Kyoto Protocol, Annex B Parties agreed to be bound by specific commitments expressed relative to a baseline year level and calculated over a five-year period, the first such commitment period being from 2008 to 2012. These quantified emission limitation and reduction commitments (QELRCs) were used to calculate an assigned amount for each Annex B Party, which each agreed not to exceed over the course of the commitment period.[65] Each Annex

---

[63]   Fn.2 (Manea) 307–8.
[64]   As in the case, for instance, of being 'equal to one metric tonne of carbon dioxide equivalent, calculated using global warming potentials …'.
[65]   Fn.61 (Article 3 KP).

B Party's assigned amount was divided into assigned amount units (AAUs), defined in the modalities, rules and guidelines for emissions trading under Article 17 of the Kyoto Protocol,[66] as being '... equal to one metric tonne of carbon dioxide equivalent, calculated using global warming potentials defined by decision 2/CP.3 or as subsequently revised in accordance with Article 5'.[67] Also set out in that decision, were the modalities for the accounting of assigned amounts[68] and paragraph 13 thereof provides for retirement of units to demonstrate compliance.[69]

The idea that all units are of equal value, namely one tonne carbon dioxide equivalent GHG (CO_2e), warrants closer consideration. In the centralized structure under the Clean Development Mechanism (CDM), for instance, all certified emission reductions (CERs) are issued by the CDM Executive Board (CDMEB) once it is satisfied that its requirements (such as a recognized methodology being applied, the project proposal being acceptable, monitoring and reporting being appropriate, and there having been verification and certification of the outcomes) are met. These requirements have been evolving over time.[70] Thus, even though there have been issues with the CDM and the CDMEB,[71] the fact that CERs all emanate from the same entity, presumably applying its criteria on a consistent basis, provides a modicum of comfort that they are all of the same value as allocated.

In the case of the Annex B Parties' assigned amounts and AAUs, from one perspective it is appropriate that all are treated as equal, since it does not matter where the GHGs are reduced, a tonne reduced in one jurisdiction would equal a tonne reduced in another. However, the issue of hot air[72] was an early signal that there were differences: not in the sense that a tonne in one location did not equal a tonne in another location, but between the actions (or lack thereof) to achieve the reduction of that tonne. Putting it another way, while a tonne is still equal to a tonne, the question is whether all AAUs are equal to a tonne mitigated? The AAUs allocated were the Party's permitted emissions

---

[66] Decision 11/CMP.1, FCCC/KP/CMP/2005/8/Add.2, 17 <http://unfccc.int/resource/docs/2005/cmp1/eng/08a02.pdf#page=17> accessed 06/06/17.

[67] Ibid, paragraph 3; similar definitions are set out for the other tradable units, certified emission reductions (CERs) under the Clean Development Mechanism, emission reduction units (ERUs) under joint implementation and removal units (RMUs) from land use, land-use change and forestry activities (LULUCF).

[68] Decision 13/CMP.1, FCCC/KP/CMP/2005/8/Add.2, 23 <http://unfccc.int/resource/docs/2005/cmp1/eng/08a02.pdf#page=23> accessed 06/06/17.

[69] Ibid, 27.

[70] See, for instance: World Bank, Ecofys, *State and Trends of Carbon Pricing 2014* (Washington, DC, 2014) 40.

[71] Discussed in Chapter 2.

[72] See Chapter 2 above and preceding section of this chapter.

for the commitment period. In theory, if the Party's emissions were less, then the surplus could be sold, if emissions exceeded the permitted amount, more would need to be acquired. Unlike the CDM, however, determination of the permitted level, monitoring and reporting actual emissions, verification, and how the reductions were actually achieved were all decentralized to the Parties individually and not necessarily consistently. Furthermore, differing levels of ambition suggest that each AAU of a more ambitious Party – that is, one that committed to a higher percentage emission reduction – would (in theory, at least) equate to a higher level of emission reduction and thus be more environmentally valuable. However, as noted earlier, emission allowances are '… first and foremost accounting units …' with two determining features being that they represent an entitlement to release a certain quantity of GHG emissions into the atmosphere (namely, one tonne CO_2e each); and they are transferable under certain established conditions.[73]

The fact that this has not become more of an issue might be ascribed to the limited AAU trading that took place under the Kyoto Protocol.[74] Whether or not this is the case, it is submitted that the only realistic way for international emissions trading (IET) under the Kyoto Protocol to have been effective would have been for Annex B Parties to establish domestic markets, thus engaging the private sector but, with notable exception of the EU, this did not happen to a significant degree.[75] As mentioned in Chapter 3, the World Bank observed in 2014, looking back on IET it cannot be assumed sovereign governments will use the market, and trading will primarily be driven by the private sector, which needs to be given an explicit role.[76]

The Kyoto Protocol evidences what has been described as a top-down governance model.[77] All units were defined as having a value of one tonne CO_2-eq

---

[73] Fn.13 (Wemaere et al) 37.

[74] The bulk of trading has been in EUAs under the EUETS and in CERs, see: fn.70 (World Bank 2014).

[75] Notwithstanding the finding of Shishlov et al. concerning compliance by the remainder 36 countries with commitments under the Kyoto Protocol: Igor Shishlov, Romain Morel, and Valentin Bellassen, 'Compliance of the Parties to the Kyoto Protocol in the First Commitment Period' (2016) 16(6) *Climate Policy* 768; also see: Michael Grubb, 'Full Legal Compliance with the Kyoto Protocol's First Commitment Period – Some Lessons' (2016) 16:6 *Climate Policy* 673.

[76] Fn.70 (World Bank 2014) 44.

[77] See, for instance: Annalisa Savaresi, 'The Paris Agreement: a New Beginning?' (2016) 34(1) *Journal of Energy & Natural Resources Law* 16, 20. Reliance on a top-down model of targets and timetables has been described as one of the three fatal flaws in the Kyoto Protocol that render it a dead-end, rather than a foundation for progress: Robert O. Keohane and Michael Oppenheimer, 'Paris: Beyond the Climate Dead End through Pledge and Review?' (2016) 4(3) *Policy and Governance* 142.

GHG. These units were either allocated, as in the case of AAUs, or issued, as with, for example, CERs. In this sense, they were all centrally sourced. In the case of the project credits, only when the projects and their outcomes had been validated and verified as having reached the required standards, were their outputs recognized and CERs issued. Thus, the top-down model operated on the basis of what might be described as unitary homogeneity. All mitigation actions had to reach the same standard so that units derived from them could be of equal value.

2.   Paris Agreement and Heterogeneity

The Paris Agreement moves away from the unitary homogeneity of the Kyoto Protocol, by recognizing that jurisdictions will take different approaches that will have different outcomes. This has been ascribed to a general trend away from specific categories differentiating parties in terms of commitments towards self-differentiation, in response to the continuing demands by developed countries for developing countries to take on commitments and developing countries continuing resistance.[78] The deal struck in Paris '… allows parties to define their own commitments, tailor these to their national circumstances, capacities, and constraints, and thus differentiate themselves from each other'.[79] The Paris Agreement '… establishes a new paradigm in international climate policy. While the Kyoto Protocol was essentially based on the so-called "targets & timetables" the Paris Agreement is based on the so-called "pledge & review" paradigm'.[80] The differences in responsibility for and in actual capacity to address climate change are implicit to this approach and evidenced through the nationally determined contributions (NDCs) that parties to the Paris Agreement have lodged. There is a considerable amount of variation in the levels of ambition disclosed, types of contributions, and target years or periods.[81]

---

[78]   Daniel Bodansky, Jutta Brunnée, and Lavanya Rajamani, *International Climate Change Law* (1st edn., Oxford University Press, 2017) 29.

[79]   Ibid.

[80]   Martin Cames et al., 'International Market Mechanisms after Paris', Discussion Paper, November 2016, German Emissions Trading Authority (DEHSt) for German Environment Agency, 7 <https://newclimate.org/2016/11/17/international-market -mechanisms-after-paris/> accessed 14/05/17.

[81]   Ibid, 15; also see: Lambert Schneider et al., 'Robust Accounting of International Transfers under Article 6 of the Paris Agreement' Discussion Paper, September 2017, German Emissions Trading Authority (DEHSt) for German Environment Agency <https://www.dehst.de/SharedDocs/downloads/EN/project-mechanisms/Differences _and_commonalities_paris_agreement_discussion_paper_28092017.pdf?__blob= publicationFile&v=2> accessed 29/09/17.

Inevitably, the mechanisms applied by different jurisdictions will vary greatly as well, including varied and diverse pricing mechanisms. The introduction by the Paris Agreement of internationally transferred mitigation outcomes (ITMOs) ties the units traded to the physical results of the mitigation actions taken.[82] This raises a number of issues, such as an appropriate measuring unit, accounting unit, how they should be represented (e.g., by a certificate), whether they could support a secondary market, whether they would all be equal and fungible, and, most importantly, what exactly is meant by a mitigation outcome.[83] Notwithstanding the similarities between the Mitigation and Sustainable Development Mechanism, introduced in Article 6, paragraphs 4–7, and the CDM under Article 12 of the Kyoto Protocol,[84] the concept of mitigation outcomes introduced by the Paris Agreement is a fundamental departure from the Kyoto Protocol concept of centrally sourced and allocated units of equal value. Nonetheless, there are many issues yet to be addressed including, for example, accounting for these diverse mitigation outcomes and how environmental integrity will be assured.

Key aspects flagged in relation to accounting under this new approach include quantifying mitigation targets and progress towards them; quantifying mitigation outcomes; avoiding double counting of reductions; accommodating different metrics for outcomes and targets; accounting for time period factor variations in outcomes and targets; and other factors affecting outcomes (e.g., non-permanence).[85] In relation to environmental integrity, in the context of international transfers of mitigation outcomes, there is support for the view this means that the transfer does not result in an increase in global aggregate emissions.[86] Four factors identified as influencing it are the robustness of accounting for international transfers; the quality of the units, which in turn depends on cap setting and monitoring; the ambition and scope of the transferring country's mitigation target; and incentives or disincentives for future mitigation action.[87] More jurisdictions have begun to develop emissions trading as part of domestic measures to achieve GHG reductions: '... 2015–2016 witnessed an increasing number of governments using or actively considering carbon pricing as an instrument to meet their emission reduction pledges and

---

[82] Thus highlighting the environmental policy reason for that trading.
[83] Fn.80 (Cames et al.) 12.
[84] Ibid, 17: see table of similarities.
[85] Fn.81 (Schneider et al.) 18–19.
[86] Lambert Schneider and Stephanie La Hoz Theuer, 'Environmental Integrity of International Carbon Market Mechanisms under the Paris Agreement' (2019) 19(3) *Climate Policy* 386. This and other issues related to ITMOs are discussed further in following chapters.
[87] Ibid, 389–92.

a growing number of companies engaging in this topic'.[88] More frequently, they are engaging in discussions aimed at facilitating inter-jurisdictional trading, for example, by linking with each other. The more this happens, the more apparent it will become that such schemes do not all generate equivalent outcomes, thus necessitating consideration of different approaches.

SUMMATION

The purpose of this chapter, together with the other chapters in this Part, is to establish a foundation for the introduction of the market proposal, the subject of this book. Analysis in this chapter of the nature of what is traded in emissions trading schemes leads to a conclusion, it is argued, that it is not the nature but the value, in mitigation terms, of what is traded that needs to be the focus for international policymakers and lawyers, in the transition to the new regime under the Paris Agreement. Valuing mitigation outcomes will be essential for establishing cooperative approaches involving international transfers under Article 6 thereof, given the variations in measures by which they might be generated. Addressing this issue is fundamental to the market proposed by this book, which also seeks to address the compartmentalization issue through the governance structure proposed. The intention is that the carbon market be treated more as other financial markets are, through that structure, allowing for greater synergies that might also engender re-engagement of the private sector.

To properly define what emission allowances are, they need to be seen in the context in which they are generated, that is, the overall scheme and its environmental purpose. Under the Paris Agreement, what can be transferred internationally is a mitigation outcome, which it is posited, means the physical benefit afforded by the emission allowance (or other unit, such as a credit) under the scheme. That is, the actual amount of GHG emission reduction that can be ascribed to a unit under that scheme. The conclusion is that just as, if not more importantly than defining the units traded in emissions trading schemes, it is their value that should be considered under Article 6. Building from these foundations, the next chapter examines what is the 'carbon market' in this new environment. This leads, in following chapters, to consideration of the infrastructure, technical, administrative, and legal frameworks, necessary to accommodate international transfers of mitigation outcomes as part of an effective and efficient mechanism of emission mitigation, within the terms of the Paris Agreement.

---

[88]   World Bank, Ecofys and Vivid Economics, *State and Trends of Carbon Pricing 2016* (Washington, DC, 2016) 28.

5. Carbon market diversity and reasons to connect

This final chapter of Part II takes another perspective on the carbon market. Having considered the interaction of the dual functions of the carbon market, as both environmental policy measure and financial trading market, from a macro-perspective in Chapter 3, and then at a more granular level in Chapter 4, both in terms of how it has evolved, this chapter examines the carbon market as it stands now and its future direction. It begins by profiling the diverse elements that might be considered to constitute the carbon market, before canvassing the rationale for connecting these diverse elements. This touches on the economic arguments, but relies on the academic literature on the subject. Mechanisms that could be applied to achieve connections between the constituent elements, in terms of emission trading schemes (ETSs), are surveyed in the final section of the chapter.

THE CARBON MARKET IN PROFILE

The carbon market can either be considered broadly as encompassing all carbon pricing (as outlined in Chapter 1), including not just emissions allowance trading schemes and project-generated credits, but also carbon taxes (credits for which might be traded) and tradable renewable energy certificates; or on the other hand, it might be construed narrowly as limited to, say, just emission allowance trading schemes. It might be categorized also according to a number of different criteria including the level at which it operates, for example, international, regional, national, subnational, municipal/local, or even on a sectoral basis, or as including internal carbon pricing applied by corporations. Alternatively, it might be classified by the legal basis on which it operates, that is, on a legal compliance basis, or on a voluntary basis; or even by the nature of the instrument traded, that is whether as allowances as part of a cap-and-trade scheme, or credits generated on a project-basis.

Taking an overall perspective, carbon market activities present a mixed picture and, at the international level, are in a state of flux. Article 17 of the Kyoto Protocol provided for international emissions trading (IET) in both allowances (assigned amount units (AAUs)) and a number of project-generated credits (including, for instance, certified emissions units (CERs), emission reduction units (ERUs)), which has occurred over the first commitment period from 2008–12. However, the Doha Amendment[1] to the Kyoto Protocol, providing for a second commitment period from 2013–20, is yet to receive the requisite number of letters of acceptance,[2] and thus is yet to take effect. As such, trading at this level is greatly reduced, although it has not discontinued entirely.[3] It is difficult to see any change in the current volume of trading activity at this level until the second Kyoto commitment period commences, or parties to the Paris Agreement[4] start to engage in cooperative approaches pursuant to Article 6 thereof. Nevertheless, at the same time, but in relation to emissions trading schemes at a regional level, the European Union Emission Trading Scheme (EUETS) continues and its framework for phase 4, from 2021–30, inter alia, is aiming to facilitate achievement of a 43 per cent reduction in greenhouse gas (GHG) emissions from covered sectors.[5] Further, at the national and subnational levels, many other ETSs have either already been implemented or are scheduled to be implemented,[6] or are under consideration.[7]

---

[1]   Doha Amendment to the Kyoto Protocol, Doha, 8 December 2012, C.N.718.2012. TREATIES-XXVII.7.c (Depositary Notification) <https://treaties.un.org/doc/Treaties/2012/12/20121217%2011-40%20AM/CN.718.2012.pdf> accessed 06/03/19.

[2]   As at 6 January 2020, 136 of the required 144 letters of acceptance had been deposited <https://unfccc.int/process/the-kyoto-protocol/the-doha-amendment> accessed 22/01/20.

[3]   UNFCCC SBI 49: Report of the administrator of the international transaction log under the Kyoto Protocol, 26 October 2018 <https://unfccc.int/sites/default/files/resource/sbi2018_inf10.pdf> accessed 06/03/19.

[4]   Report of the Conference of the Parties on its twenty-first session, held in Paris from 30 November to 13 December 2015. Addendum. FCCC/CP/2015/10/Add.1, 29 January 2016 <http://unfccc.int/resource/docs/2015/cop21/eng/10a01.pdf> accessed 13/03/17.

[5]   See, for instance: European Commission, 'Report from the Commission to the European Parliament and the Council, Report on the functioning of the European carbon market', Brussels, 17.12.2018 COM(2018) 842 final <https://eur-lex.europa.eu/legal-content/EN/TXT/PDF/?uri=CELEX:52018DC0842&from=EN> accessed 06/03/19.

[6]   World Bank and Ecofys, *State and Trends of Carbon Pricing 2018* (Washington, DC, 2018) <www.worldbank.org> accessed 12/08/18. *National ETSs:* Australia, Austria, Belgium, Bulgaria, China, Croatia, Cyprus, Czech Republic, Germany, Greece, Hungary, Italy, Kazakhstan, Lithuania, Luxembourg, Malta, the Netherlands, New Zealand, the Republic of Korea, Romania, and Slovakia. *Both national ETSs and carbon taxes:* Denmark, Estonia, Finland, France, Iceland, Ireland, Latvia, Liechtenstein, Norway, Poland, Portugal, Slovenia, Spain, Sweden, Switzerland, and the United Kingdom. *Subnational ETSs:* Beijing, California, Chongqing, Connecticut, Delaware, Fujian, Guangdong, Hubei, Maine, Maryland, Massachusetts, New Hampshire, New York, Ontario, Québec, Rhode Island, Saitama, Shanghai, Shenzhen, Tianjin, Tokyo, Vermont, and Washington State. *Both subnational ETSs and carbon taxes:* Alberta and British Columbia.

[7]   Ibid. *National ETS or carbon tax:* Brazil, Canada, Chile (ETS), Colombia (ETS), Côte d'Ivoire, Japan (ETS), Mexico (ETS), the Netherlands (carbon tax), Thailand,

In terms of project-generated credits, at the international level, even though the Clean Development Mechanism (CDM) is not directly dependent for its continued operation on the second Kyoto commitment period,[8] CDM activity has been waning since reaching a peak at the end of 2012. The absence of demand from parties with commitments under a second commitment period, as well as demand from the EUETS being cut,[9] is compounded by the fact that the status of certified emission reductions (CERs) as mitigation outcomes under Article 6, paragraph 2, or for the purpose of the mechanism under Article 6, paragraph 4, Paris Agreement remains unclear.[10] The drop off in CDM is evidenced by the fact that just three projects entered the validation process in 2017, this being last such activity, although there are still projects further along in the registration process.[11]

Notwithstanding, developments at a sectoral level in relation to aviation and shipping may eventually provide a fillip for project-generated credits. While tangible developments in relation to shipping are yet to crystallize, the International Civil Aviation Organization (ICAO) has resolved to develop a global market-based measure scheme for international aviation in the form of the Carbon Offsetting and Reduction Scheme for International Aviation (CORSIA).[12] Participation of ICAO member states in the pilot phase (2021–23) and first phase (2024–26) is voluntary, but from the second phase commencing 2027, all states with an individual share of international aviation

---

Turkey, Ukraine (ETS), and Vietnam. *Subnational ETS or carbon tax:* Catalonia, Manitoba, New Brunswick, Newfoundland and Labrador, New Jersey, Northwest Territories, Nova Scotia, Oregon, Prince Edward Island, Rio de Janeiro, São Paolo, Saskatchewan, Taiwan, China, and Virginia.

8   Although obviously it is dependent on the second commitment period to the extent that commitments thereunder may generate demand for CERs.

9   The EU has a domestic emission reduction target that does not envisage the use of international credits after 2020 <https://ec.europa.eu/clima/policies/ets/credits_en> accessed 06/03/19.

10   See, for instance, the draft definition of ITMOs at paragraph VI, C: UNFCCC SBSTA 48-2: Draft Text on SBSTA 48-2 agenda item 12(a) Matters relating to Article 6 of the Paris Agreement: Guidance on cooperative approaches referred to in Article 6, paragraph 2, of the Paris Agreement Version 1 of 9 September 02:00 hrs - corrected version   <https://unfccc.int/sites/default/files/resource/sbsta48.2_12a_DT_corr.pdf> accessed 29/10/18.

11   See <https://cdm.unfccc.int/Statistics/Public/CDMinsights/index.html> accessed 05/03/19.

12   International Civil Aviation Organisation, 'Resolution A39-3: Consolidated statement of continuing ICAO policies and practices related to environmental protection – Global Market-based Measure (MBM) scheme', Assembly 39th Session, October 2016, Paragraph 5 <https://www.icao.int/environmental-protection/CORSIA/Documents/Resolution_A39_3.pdf> accessed 06/03/19.

activity above a threshold level are included.[13] CORSIA will use emission units that meet specified Emission Unit Criteria developed on the advice of a technical advisory body, which should promote compatibility with future relevant decisions under the Paris Agreement.[14]

Apart from project-generated credits under the Kyoto Protocol, there are also a number of voluntary standards schemes[15] under which projects generate carbon market offsets used by corporations and other entities, that otherwise do not have compliance obligations, to offset their emissions voluntarily. These voluntary standards differ according to project activities and types allowed, project locations and the regulations to which the projects must adhere, but generally all the voluntary standards require the offsets to be real, additional, measurable, and verifiable.[16] While this is a comparatively small part of the carbon market, it is not negligible and claims to have offset over 437 million tonnes carbon dioxide equivalent (tCO_2e) GHG since 2005, as well as providing sustainable development co-benefits such as supporting local economies through job training and creation, preserving watersheds that supply clean water, or safeguarding biodiversity.[17]

Since the Paris Agreement in 2015, it seems there is a clear trend towards greater implementation of carbon pricing initiatives, at various levels of government around the world.[18] The World Bank has reported, for instance, that 88 Parties to the Paris Agreement, accounting for 56 per cent of global GHG emissions, have indicated they are planning or considering use of carbon pricing and/or market mechanisms.[19] It noted also that over 1300 companies globally are using or planning to use internal carbon pricing in 2018–19.[20] In the broadest sense, then, the carbon market can be seen as a diverse, heteroge-

---

[13] Ibid, paragraph 9(e), other than least developed countries, small island developing states and landlocked developing countries, unless they join voluntarily.

[14] Ibid, paragraph 20(c)–(e).

[15] Verified Carbon Standard: <https://www.verra.org>; Gold Standard: <https://www.goldstandard.org>; Plan Vivo: <www.planvivo.org>; Climate Action Reserve: <www.climateactionreserve.org>.

[16] Kelley Hamrick and Melissa Gallant, 'Voluntary Carbon Markets Insights: 2018 Outlook and Trends', 2018 *Ecosystem Marketplace, A Forest Trends Initiative* <https://www.forest-trends.org/wp-content/uploads/2018/09/VCM-Q1-Report_Full-Version-2.pdf> accessed 05/03/19.

[17] Ibid.

[18] World Bank, Ecofys and Vivid Economics, *State and Trends of Carbon Pricing 2016* (Washington, DC, 2016) 11 <https://openknowledge.worldbank.org/bitstream/handle/10986/25160/9781464810015.pdf?sequence=7&isAllowed=y> accessed 28/06/17.

[19] Fn.6 (World Bank 2018). It is noted that this number is less than in preceding World Bank reports, but substantial all the same.

[20] Ibid.

neous collection of different types of pricing mechanisms, being implemented, or scheduled to be implemented, by a range of different levels of government, and other stakeholders, encompassing both voluntary commitments and legally binding obligations.

There is also a wide variation in the carbon prices that apply across these mechanisms. While prices appear to be increasing slightly year-on-year, the current price trajectories are insufficient to stimulate emission reductions in line with the Paris Agreement temperature goals.[21] Prices range from US$140 per tCO_2e (carbon tax in Sweden) to less than US$1 for the carbon taxes in Mexico, Poland, and Ukraine, while ETS prices range from US$23 in Alberta, Canada, to US$2 in the Hubei and Guangdong pilot ETS schemes in the People's Republic of China (PRC).[22] The High Level Commission on Carbon Prices established pursuant to the 22nd Conference of Parties to the UNFCCC (COP22), has concluded that while countries may choose different instruments to implement their carbon policies, depending on national and local circumstances and support received, the explicit carbon price level consistent with achieving the Paris Agreement temperature target is at least US$40–80 per tCO_2e by 2020 and US$50–100 per tCO_2e by 2030.[23] As noted in the Stern Review: 'If the carbon price across countries is not broadly similar, there will be unexploited opportunities to abate an extra tonne of GHG more cheaply in one country compared with another, so the overall cost of abatement will be higher'.[24] Thus, an urgent challenge for international collective action, Stern states, is a broadly similar global carbon price.[25] The currently diverse range of carbon prices across the various mechanisms might be seen, therefore, as not only being insufficient, but also inefficient.

An additional inefficiency, it is argued, arises from the differences that exist not just in prices, but also across many design and other aspects of pricing mechanisms. The mechanisms, even those of the same type, such as ETSs, will differ from each other, at least to some degree, in terms of design, the rules and standards they apply, the extent of their coverage, how they are implemented, the framework of policies and ambition of which they form part, and in the legal, economic, social, and political context of the jurisdiction in

---

[21] Ibid, 27.

[22] Ibid, 21.

[23] World Bank Group, Networked Carbon Markets: Mitigation Action Assessment Protocol, 2016, World Bank, Washington, DC. © World Bank. <https://openknowledge .worldbank.org/bitstream/handle/10986/25371/110153-WP-P161139-PUBLIC -MAAPMay.pdf?sequence=1&isAllowed=y> accessed 27/02/18.

[24] Nicholas Stern, *The Economics of Climate Change: The Stern Review* (Cambridge University Press, 2007) 532.

[25] Ibid.

which they exist.[26] For instance, share of allowances not provided for free (in other words, that must be acquired at auction or otherwise) can range from 100 per cent under the Regional Greenhouse Gas Initiative (RGGI) covering 165 power generators across nine north-eastern US states, to 0 per cent under the Korean ETS which covers 599 entities from a number of business sectors; or the percentage of emissions covered, which ranges from 85 per cent under the Western Climate Initiative (WCI) over one US state and four Canadian provinces, to 68 per cent for the Korean ETS, 52 per cent for the NZ ETS, 45 per cent for the EUETS down to 20 per cent for RGGI.[27] Other such differences are discussed in the following section.

CONNECTING DIVERSE EMISSIONS TRADING SCHEMES

The two issues of first, increasing prevalence and diversity of carbon pricing mechanisms in jurisdictions around the world, and second, the corresponding complexity of, yet need to value comparatively, the units of measure traded in these heterogeneous mechanisms (to facilitate international transfers envisaged by the Paris Agreement), coalesce in the question of how to connect such mechanisms and, in particular, how to connect ETSs. This question is integral to the market proposed by this book, which envisages a future where there may be trading between the heterogeneous ETSs and other pricing mechanisms that are being developed. This proposal proceeds on the basis that trading across schemes and jurisdictions is desirable for a number of reasons, outlined in this and following sections.

At its broadest, the proposal posited refers to a market in which any form of mitigation outcome could be traded, implying that there would be a valid and generally accepted methodology for valuing and comparing various forms of mitigation outcome. For the sake of simplicity and clarity, the proposal outlined is couched in terms of a market (initially) between ETSs, or put another way, a market that connects those ETSs. Much has been written and discussed about how various jurisdictions' ETSs might be better linked to one another,[28]

---

[26]　See, by way of example of the variety: International Carbon Action Partnership, *Emissions Trading Worldwide: Status Report 2016*. Berlin <https://icapcarbonaction .com>.

[27]　International Carbon Action Partnership (ICAP), *Emissions Trading Worldwide: Status Report 2018*, ICAP, Berlin. <https://icapcarbonaction.com/en/?option=com _attach&task=download&id=547> accessed 07/03/18.

[28]　For summaries and overviews of academic research and issues relating to linking emissions trading schemes, see: Michael Mehling, 'Legal Frameworks for Linking National Emissions Trading Systems' in C. Carlarne, K. Gray, and R. Tarasofsky

but there are only a few instances where such linking has actually taken place.[29] Three examples are first, the link that was established in 2004 between the EUETS and the Kyoto Protocol project mechanisms, noting that this specifically excluded the use of project credits generated by nuclear facilities, from land use, land use change, and forestry activities, and from large hydro projects that did not conform to the criteria specified by the World Commission on Dams;[30] second, the agreement of September 2013 between the US State of California and the Canadian Province of Québec to harmonize and integrate cap-and-trade programs;[31] and the agreement between the European Union and the Swiss Confederation to link their respective ETSs.[32]

This section proceeds by reviewing a selection of the literature and by considering the rationale put, in general, in support of connecting emissions markets. The term 'connecting' is used as a generic expression inclusive of both linking and networking, since the reasons in support cited in the literature,

---

(eds), *Oxford Handbook of International Climate Change Law* (Oxford University Press, 2016) 261; Aki Kachi et al., *Linking Emissions Trading Systems: A Summary of Current Research*, January 2015, ICAP <https://icapcarbonaction.com/en/?option= com_attach&task=download&id=241> accessed 06/09/17; Intergovernmental Panel on Climate Change (IPCC), 'International Cooperation: Agreements and Instruments' in Climate Change 2014: Mitigation of Climate Change. Contribution of Working Group III to the Fifth Assessment Report of the Intergovernmental Panel on Climate Change, [Edenhofer, O., et al. (eds.)]. Cambridge University Press <https://www.ipcc.ch/pdf/ assessment-report/ar5/wg3/ipcc_wg3_ar5_chapter13.pdf> accessed 31/07/17.

[29]   As opposed to being 'networked' with each other, which is presently only conceptual, so there are no instances of networking that might be cited by way of example. For 'linking', on the other hand, the arrangement between the ETSs of the US state of California and the Canadian province of Quebec (and Ontario from January 2018, until July 2018 when regulation 386/18 terminated cap and trade regulation 144/16) is an example. Others are the EU Linking Directive with the Kyoto Protocol mechanisms, the EU-Switzerland ETS link (yet to be ratified) and, in Japan, the link between the Tokyo Metropolitan Government cap-and-trade scheme and the Saitama Prefecture ETS; and prior to joining the EUETS in 2007, Norway had a one-way link to the EUETS from 2005.

[30]   Directive 2004/101/EC of the European Parliament and of the Council of 27 October 2004 amending Directive 2003/87/EC establishing a scheme for greenhouse gas emission allowance trading within the Community, in respect of the Kyoto Protocol's project mechanisms, OJ L 338, 13.11.2004, 18–23.

[31]   Agreement between California Air Resources Board and the Government of Quebec, Concerning the Harmonisation and Integration of Cap-And-Trade Programs for Reducing Greenhouse Gas Emissions, 27 September 2013 <https://www.arb .ca.gov/cc/capandtrade/linkage/ca_quebec_linking_agreement_english.pdf> accessed 06/03/18.

[32]   Agreement between the European Union and the Swiss Confederation on the linking of their greenhouse gas emissions trading systems. OJ L 322, 7.12.2017, 3–26. Agreement finally took effect 1 January 2020.

for the most part, apply equally to both.[33] All the same, before proceeding, it is helpful to make a brief introduction of linking and networking. One description of linking is: '… emissions trading systems are linked if a participant in one system can use a carbon unit issued under another system to meet compliance obligations … units are considered fungible, or equivalent for compliance purposes …'[34] For this to be the case, linking entails a certain level of agreement between jurisdictions. As a precondition, they would need to consider, at least, the compatibility of their ETSs, but also, one might expect, comparability of their economies and emissions profiles. Once linked, there might also be expected some degree of convergence between the systems. An indirect link may also occur where two or more jurisdictions accept project-generated credits from the same source, an obvious example being jurisdictions accepting CERs generated by CDM projects registered under the Kyoto Protocol.[35]

Networking would be a more flexible arrangement, entailing less need for convergence between the jurisdictions and their respective ETSs, because it places a value on the differences. As such, it requires no harmonizing or integrating of physical infrastructure, laws, policies, administration, or other elements from the respective jurisdictions, just agreement as to the connection via which the transaction may proceed. Thus, having arrived at values for the respective units, a conversion factor can be derived, which is used in the transfer transaction between the networked jurisdictions' ETS systems.[36]

1.   The Rationale for Connecting

Economic, political, and environmental arguments have been advanced that trading across schemes and jurisdictions is desirable for a number of reasons, broadly including:

- because it can foster larger, deeper, and more liquid markets, that are less susceptible to manipulation and that more effectively price carbon emissions;
- greater efficiency and scale might be achieved in those markets;

---

[33] Differences between the two mechanisms are elaborated, as are fuller definitions, in the following section.

[34] Fn.28 (Mehling) 261.

[35] As would be the case with the EUETS and another ETS that accepted Kyoto Protocol project credits.

[36] See Justin Macinante, 'Networking Carbon Markets – Key Elements of the Process', 2016, World Bank Group Climate Change <http://pubdocs.worldbank.org/en/424831476453674939/1700504-Networking-Carbon-Markets-Web.pdf> accessed 01/03/18.

- cross-jurisdictional trading could generate a more globally consistent, stable carbon price by reducing price volatility;
- trading across jurisdictions would reduce the risk of carbon leakage;
- politically it might demonstrate leadership, allowing pressure to be exerted on free-riding nations;
- it may offer domestic support for emissions trading, indicating positive momentum;
- administrative costs would be less; and
- it would be more encouraging of investment in climate finance.[37]

Four arguments are usually stressed, by the economic literature, in favour of linking, being that it: affords higher cost-efficiency through a larger number of mitigation options; provides a more robust price signal; reduces distortions, through converging carbon prices; and increases market liquidity due to more market participants.[38] Linking can make an ETS a viable policy option, which may not be the case without linking, by reducing the cost of achieving the combined emissions cap of the linked ETSs, and by increasing the size of the market, thereby generating more liquidity, reducing the market power for larger participants and increasing the availability of more financial instruments, facilitating negotiation of trades, and lowering transaction costs.[39]

Of its nature, 'linking results in an enlarged market, promising greater diversity of abatement costs and thus more efficient achievement of mitigation objectives'.[40] By promoting liquidity and reducing price volatility in the market, it should help reduce the likelihood of market manipulation and abuse.[41] It is also seen as one of the few options for meaningful collective

---

[37] Daniel M. Bodansky et al., 'Facilitating Linkage of Climate Policies through the Paris Outcome' (2016) 16(8) *Climate Policy* 956; Daniel M. Bodansky et al., 'Facilitating Linkage of Heterogeneous Regional, National, and Sub-National Climate Policies through a Future International Agreement'. Cambridge, Mass.: Harvard Project on Climate Agreements, November 2014; World Bank, Partnership for Market Readiness, *Lessons Learned from Linking Emissions Trading Systems: General Principles and Applications*, Technical Note 7, February 2014. World Bank, Washington, DC. © World Bank; also fn.28 (Mehling).

[38] Christiane Beuermann et al., 'Considering the Effects of Linking Emissions Trading Schemes, A Manual on Bilateral Linking of ETS', May 2017, German Emissions Trading Authority (DEHSt) on behalf of German Environment Agency <https://www.dehst.de/SharedDocs/downloads/EN/emissions-trading/Linking_manual.pdf?__blob=publicationFile&v=3> accessed 17/07/17.

[39] Fn.37 (World Bank, PMR, 2014) 8.

[40] Fn.28 (Mehling) 258.

[41] Ibid.

action.[42] All the same, reasons given by jurisdictions that have linked, or intended or attempted to link, are predominantly economic. For example, the EU has emphasized lower compliance costs, increased market liquidity and price stability (although it also mentions increasing global cooperation and levelling the playing field), while Switzerland and New Zealand both flag increased liquidity and greater flexibility for regulated entities – far more significant issues for these smaller economies, one might expect, compared to the EU – although California also sees linking as offering greater market liquidity and flexibility to its regulated entities.[43]

Notwithstanding the emphasis on economic benefits, other authors identify political benefits that are centred on the political signals of a common effort to address climate change, enhanced cooperation and influence, limiting competitiveness concerns, and enabling the adoption of more ambitious targets.[44] There are also administrative benefits of sharing best practices, and lowering compliance and administration costs.[45]

2. Risks of Connecting

Despite these positive aspects, the process of linking has been described as '... procedurally demanding and politically complex ...'[46] as there are potential risks. For instance, once linked, design features of one system can extend to the other, possibly compromising environmental objectives and sovereign control.[47] Furthermore, convergence of prices 'may have distributional impacts on participants and other stakeholders ... potentially resulting in substantial capital flows across borders and undermining political support for continued linkage'.[48] The legal and institutional considerations, also, 'can ultimately

---

[42]   Ibid; see also fn.28 (Kachi et al./ICAP); also fn.37 (Bodansky et al.); Matthew Ranson and Robert N. Stavins, 'Linkage of Greenhouse Gas Emissions Trading Systems: Learning from Experience' Discussion Paper ES 2013-2. Cambridge, Mass.: Harvard Project on Climate Agreements, November 2013; Judson Jaffe and Robert N. Stavins, 'Linkage of Tradable Permit Systems in International Climate Policy Architecture' Discussion Paper 2008-07, Cambridge, Mass.: Harvard Project on International Climate Agreements, September 2008.

[43]   Fn.38 (Beuermann et al./DEHSt) at 13 (Box 4).

[44]   Fn.28 (Kachi et al./ICAP) 4; also Michael Lazarus et al., (Stockholm Environment Institute) *Options and Issues for Restricted Linking of Emissions Trading Systems*, September 2015, ICAP Berlin, Germany, 6.

[45]   Ibid (Kachi et al./ICAP).

[46]   Fn.28 (Mehling) 258.

[47]   Ibid, 259.

[48]   Ibid.

undermine whether an emissions trading link becomes operational'.[49] For instance, design features such as type and stringency of the cap, respective offset crediting provisions, commitment periods, price management mechanisms such as banking and borrowing, and governance and compliance enforcement, all need to be consistent or harmonized for linking to become operational.[50] For example, the EU-Swiss linking agreement specifies essential criteria, including such matters as the cap, level of ambition, and in relation to international credits, that need to be met by the two ETSs.[51]

Others have noted that changes in the distribution of costs in each of the connected ETSs can affect the competitiveness by increasing production costs for firms that have emission intensive inputs.[52] There may also be an incentive for each ETS to make smaller reductions over time, since this should reduce the amount of compliance instruments needing to be imported from the other jurisdiction.[53] Also, each administrator loses some control over their own ETS, which may cause negative sentiment.[54] One commentator goes further, arguing that carbon markets should not be linked, as linking would only deliver greater complexity and fewer emission reductions.[55] A solution proposed to the problems identified is the introduction of a 'central carbon bank', although creating a new international institution and insulating it from political influence may be difficult.[56] Another warning, specifically in relation to the EUETS, is that any link by the EUETS with a lower cost market, with lower ambition, would mean

---

[49] Ibid; also Michael Mehling, 'Linking of Emissions Trading Schemes' in David Freestone and Charlotte Streck (eds.), *Legal Aspects of Carbon Trading: Kyoto, Copenhagen and Beyond* (Oxford University Press, 2009) 110; also fn.44 (Lazarus et al./ICAP) 7–8.

[50] M. J. Mace et al., 'Analysis of the legal and organisational issues arising in linking the EU Emissions Trading Scheme to other existing and emerging emissions trading schemes', Final Report, May 2008, Study Commissioned by the European Commission DG-Environment, Climate Change and Air, chapter 3

[51] Fn.32 (EU-Swiss Confederation Agreement) Article 2; Annex 1.

[52] Fn.37 (World Bank, PMR, 2014) 9.

[53] Ibid. Although, this will also be dependent on the terms and nature of the agreement between the connecting jurisdictions. Re similar strategic behaviour, see also: fn.37 (Bodansky et al., 2016) 958.

[54] Ibid (World Bank, PMR, 2014) 9. Again, the extent to which this is an issue will be a function of the agreement between the connecting jurisdictions.

[55] Jessica F. Green, 'Don't Link Carbon Markets' (2017) 543 *Nature* 484. This author cites the EUETS link to the CDM and the California-Quebec link as two examples that have gone wrong, but it is not clear from these examples that the cause of the problems identified has been the fact of linking. Rather, it may have exacerbated problems that are, in fact, due at least initially, to other causes such as poor design.

[56] Ibid, 486.

fewer reductions achieved domestically.[57] The basic point made, was that no other current markets are compatible with the EUETS, a further issue being the current EUETS surplus.[58]

3.   Consideration in the Literature

The study of connecting ETSs has been traced back to when jurisdictions first began considering establishing domestic emissions trading as a mechanism for mitigation, in the context of Kyoto Protocol commitments.[59] An overview of the research undertaken distinguishes three phases, being a conceptual phase, prior to the Kyoto Protocol entering into force; an instrumental phase, when there was a focus on specific conditions and mechanisms for successful linking; and a critical phase, with the concept established in the mainstream of policy discussions.[60]

Consideration given in the literature to connecting emission trading schemes has been, to date, almost exclusively on linking: 'The majority of studies on linking to date have focused on "full" bilateral linking in which compliance instruments (allowances, offset units) are fully fungible in all participating systems'.[61] As might be expected, given this focus, studies concentrate on the degree to which elements of the respective ETSs must be harmonized for linking to work. For example, according to a World Bank analysis, design features that need to be harmonized to address political concerns are the type of cap (absolute or intensity based); stringency of the cap; offset crediting provisions; commitment periods; and stringency of enforcement; while design features that need to be harmonized to protect environmental integrity or market operation are cost containment provisions and the exclusion of ex post issuance of allowances (except to new entrants).[62] The elements of the California-Quebec Linking Agreement are cited as an example, in that report, where the harmonization and integration process included regulatory harmonization; offset protocols; mutual recognition of compliance instruments;

---

[57]   Carbon Market Watch, 'Towards a Global Carbon Market – Risks of Linking the EU ETS to other carbon markets', Policy Brief, 05 May 2015 <http://carbonmarketwatch .org/towards-a-global-carbon-market-risks-of-linking-the-eu-ets-to-other-carbon -markets/> accessed 06/09/16. It was not clear from the paper why the authors assumed that the lower price market would automatically have less ambition.

[58]   Ibid.

[59]   Fn.28 (Mehling) 262.

[60]   Ibid.

[61]   Fn.28 (Kachi et al./ICAP) 10.

[62]   Fn.37 (World Bank, PMR, 2014) 16 (Box 3).

joint auctions; and common registry and auction platforms.[63] As mentioned, elements that might be potential barriers to linking, in relation to which harmonization is important, include the nature and stringency of the cap; borrowing provisions; offset provisions; and price ceilings/floors.[64] Elements where harmonization would facilitate operation of the linked system were considered to be the monitoring, reporting, and verification (MRV) systems; registry designs; compliance periods; banking provisions; and enforcement/ penalty provisions.[65]

Another study, by Mehling, considered ETS design features that would be essential to mutual compatibility of the linked systems as being, in relation to scope and timeline, the continuity of the scheme; in relation to the cap, whether it was relative or absolute; and for cost containment, price ceilings and borrowing provisions.[66] Mehling also notes 'typologies' of linkages: these may be unilateral, bilateral, multilateral, or reciprocal unilateral.[67] Others also include indirect links as a category.[68] Some studies even look beyond ETS harmonization: 'Suggested ways to facilitate linking without the full harmonisation of key ETS design elements include restrictions on traded volume and the imposition of levies, taxes or an exchange rate that establishes a different compliance value to allowances from different schemes'.[69] These authors also flag the need to contemplate further investigation of possible legal aspects of linking subnational, national, supranational, and multilateral instruments, within a larger framework.[70]

Notwithstanding the emphasis on full bilateral linking in the literature, one study examines alternatives, short of full linking, that jurisdictions could pursue to capture some of the political, economic, and environmental benefits associated with linking.[71] It reports on modelling carried out of four alternative

---

[63] Ibid, 18 (Box 4). See: fn.31 (California-Quebec Agreement) recitals; see also fn.32 (EU-Swiss Confederation Agreement).

[64] Fn.28 (Kachi et al./ICAP) 12 (Table 1). This paper also provides, in Annex A, a detailed ETS design element overview according to implications for linking in terms of political, economic and environmental considerations.

[65] Ibid.

[66] Fn.49 (Mehling) 115 (Table 5.1).

[67] Ibid, 119–22.

[68] Fn.37 (World Bank, PMR, 2014) 7. See, for example the EU Linking Directive: Directive 2004/101/EC of the European Parliament and of the Council of 27 October 2004 amending Directive 2003/87/EC establishing a scheme for greenhouse gas emission allowance trading within the Community, in respect of the Kyoto Protocol's project mechanisms, OJ L 338, 13.11.2004, 18–23.

[69] Fn.28 (Kachi et al./ICAP) 12.

[70] Ibid.

[71] Fn.44 (Lazarus et al./ICAP).

restricted linking options, namely, quotas, one-way linking, exchange rates, and discount rates. These options were compared to control situations of no linking and full linking, and analyzed with respect to four broad criteria, being the environmental benefit; economic benefit; political feasibility; and other practical and overarching considerations.[72]

The study found that restricted linking options do not achieve the full potential benefits of full linking, but do lessen some of the pitfalls. The benefits and risks of linking are not just economic, but also environmental, and among the restricted linking options considered, discount rates could increase the abatement outcome, while exchange rates could potentially increase or decrease it.[73] Overall, restricted linking could reduce, but not wholly avoid, the need for harmonization of the ETS design elements required for full linking.[74]

It was observed that the implications and feasibility of linking (either restricted or full) depend heavily on the design of ETSs, ambition of their caps, the size of the ETSs, their marginal abatement cost curves, and use of offsets.[75] While exchange rates provide full liquidity, as under full linking, they could be affected by information asymmetries and uncertainties in the rate setting. Exchange rates could also strongly affect the location, level, and cost of abatement. The study found also that they could affect the transfer payments, auctioning revenues or any co-benefits, in a similar way to full linking.[76] The overall conclusion was that exchange rates could generate environmental and economic benefits, or lead to adverse impacts, depending on how they are set.

Existing or planned linkages between sub-national, national, and regional ETSs have also been considered in the broader context of international cooperation. In its Fifth Assessment Report, the Intergovernmental Panel on Climate Change (IPCC) critically examined and evaluated the ways in which agreements and instruments for international cooperation to address climate change have been organized and implemented.[77] Climate change policy architectures were classified into three categories as strong multilateralism; harmonized national policies; and decentralized architectures and coordinated national policies, which included linked ETSs.[78] Broadly, policies could be evaluated in terms of four criteria, namely, their environmental effectiveness, or the extent

---

[72]   Ibid. The other practical and overarching considerations include things such as administrative costs, complexity, communication difficulty, and potential impact on economic resilience.

[73]   Ibid, 35.

[74]   Ibid.

[75]   Ibid. These points coincide with other studies mentioned earlier in this chapter.

[76]   Ibid.

[77]   Fn.28 (IPCC).

[78]   Ibid, 1022 (Table 13.2).

to which the policy achieves its objective of reducing the causes and impacts of climate change; aggregate economic performance, being both economic efficiency and cost effectiveness; distributional impacts, being the burden and benefit sharing across countries and across time; and institutional feasibility, which was considered in terms of participation, compliance, legitimacy, and flexibility.[79]

The IPCC found that review of unilateral and bilateral linkages demonstrated that bilateral direct linkage could reduce mitigation costs, increase credibility of the price signal and expand market size and liquidity, but also raised concerns, first, over mitigation dilution, since the linked system would only be as effective as the weakest performer and, second, that jurisdictions may be unwilling to accept carbon price increases resulting from a link.[80] Other findings reflect the issues, mentioned earlier, over compatibility of respective ETSs. The IPCC noted also that bilateral links face lengthy adoption procedures as well as legal and other constraints, while less formal arrangements, for instance, reciprocal unilateral links provide similar benefits but may be easier to implement and more flexible.[81]

MECHANISMS FOR CONNECTING

In outlining the rationale for, and the risks of, connecting ETSs, the preceding section is expressed in terms of linking. The less formal, more flexible arrangements referred to by the IPCC might well have also included the proposal advanced by this book, for networking of carbon markets. This section, therefore, elaborates what is meant by networking and how it differs from linking.

1.   Linking

Direct bilateral or multilateral linking entails a formalized arrangement between the participating jurisdictions. As a precondition, jurisdictions considering linking would need to consider, at least, the compatibility of their respective ETSs. They would then need to adapt to each other, even to converge, and once linked, it is perhaps inevitable that the economically larger will be

---

[79]   Ibid, 1009–10. Interestingly, IPCC assessment of proposed international cooperation by linking ETSs, based on the four criteria, is expressed in terms of the quality, effectiveness, or similarity of the specific national policies, thus more in terms of the parts, as opposed to the sum of the parts; it also found that there are gaps in the literature on international cooperation concerning mitigation, for instance, few comparisons exist of proposals in terms of the four criteria (at 1053).

[80]   Ibid, 1030.

[81]   Ibid.

favoured. In addition, as jurisdictions' economies and emissions profiles do not remain static over time, imbalances and changes in balances will need to be managed as an ongoing issue. For example, in relation to linking, the World Bank observes:

> The balance of environmental benefits and distribution of costs and hence, the design features, differ for each ETS. Each ETS also reflects the institutional structure, economic circumstances, culture and traditions and other characteristics of the implementing jurisdiction. Each jurisdiction has its own legislative process for implementing and amending the ETS. The economic structure and vulnerability to external competition is unique to each jurisdiction. And each jurisdiction has its own currency and language(s).[82]

These differences are important to jurisdictions and a challenge to linking, which requires a level of harmonization that implies the need for compromises in respective ETS designs and perhaps other jurisdiction specific considerations. There are risks of reducing ambition, of 'the perceived loss of regulatory autonomy' and 'of unequal institutional and technical capacities' and 'competing domestic agendas, which may need to be reconciled' in any particular instance.[83] Successful linking requires matching jurisdictions with compatible ETS designs and policy objectives, and finding the right level at which to engage.[84]

As the decision to link is a voluntary decision on the part of each linking jurisdiction, essential requirements include a political decision by each that the benefits outweigh the risks; sufficient compatibility in the ETS design elements; arrangements to maintain compatibility and consistency over time in the face of economic and other developments; and a legal agreement to cement and implement the link.[85] As such, it has been noted: 'To-date, bilateral links between ETS are rare, perhaps due to the limited number of systems or the low probability of finding two jurisdictions where, at the same time, the political leaders appreciate the benefits of a bilateral link'.[86]

---

[82]   Fn.37 (World Bank, PMR, 2014) 11.
[83]   Fn.44 (Lazarus et al./ICAP) 4.
[84]   Ibid.
[85]   Fn.37 (World Bank, PMR, 2014) 11, Box 1.
[86]   Ibid, 12.

2.   Networking

The concept of networked carbon markets (NCM)[87] is an initiative taken forward by the World Bank.[88] NCM requires '... (i) a transparent, reliable, efficient approach to providing the information needed to determine the relative climate change mitigation value of units to be traded internationally, (ii) infrastructure to assist jurisdictions to manage carbon market-related risks and track international exchanges'.[89] The infrastructure envisaged in this early World Bank framing of the concept comprises an international carbon asset reserve (ICAR), to support and facilitate carbon market-related functions, and an international settlement platform (ISP), to track cross-border trades and possible clearinghouse functions.[90] Thus, NCM as introduced by the World Bank comprises three elements: a mechanism to measure mitigation value of mitigation outcomes; the ICAR; and an ISP.

Although still only conceptual, there being no concrete example of net-working having been implemented, it is beginning to be acknowledged in the literature. Two such instances are interesting as they highlight a fundamental difference between networking and linking. The study on restricted linking, described in the preceding section,[91] (the 'first reference') refers to NCM in the context of policy questions, such as who sets exchange rates and how; whether rates would be fixed or floating; and how they could be updated while retain-ing the integrity of allowance markets.[92] The authors note that it is unclear how rate setting would work in practice, and also that unlike other products and services and the currencies used for their exchange, emission allowances have no value outside the markets created by regulators.[93]

The second reference describes NCM as 'Probably the most comprehensive exploration to date of a hub-based architecture for carbon trading systems employing exchange rates ...'[94] The author, Mehling, is discussing an alter-native to the fungibility of traded units being based on a guiding principle of full system compatibility and equal unit value. The alternative is reliance on

---

[87]   Also (mis)described as 'heterogeneous linking' or 'soft linking'.

[88]   See: <http://www.worldbank.org/en/topic/climatechange/brief/globally-network ed-carbon-markets>; also, in particular, fn.36 (Macinante).

[89]   World Bank, (2015) Overview of Networked Carbon Markets: <http://pubdocs .worldbank.org/en/450811484257514457/Overview-of-Networked-Carbon-Markets .pdf> accessed 25/07/17.

[90]   Ibid.

[91]   Fn.44 (Lazarus et al./ICAP).

[92]   Ibid, 14.

[93]   Ibid, 28.

[94]   Fn.28 (Mehling) 276.

the metric of comparability.[95] Thus, rather than 'alignment of design features ... participation in a common market could be based on adherence to a set of minimum conditions' for design requirements, then by assessing the design quality, using discount factors, ratios, or exchange rates, to adjust mitigation values of units.[96] Mehling notes that such mitigation value rating could even enable linkages across policies other than ETS, such as carbon taxes, or even performance standards.[97] A centralized administration for such rate setting 'would significantly increase transparency and lower transaction costs'.[98] Hence: 'Jurisdictions that have introduced carbon markets could voluntarily "opt in" if they agree to have their traded units (or "carbon asset classes") rated for their "Mitigation Value" (MV) by independent private rating agencies on the basis of a standardised process and formula'.[99]

The fundamental difference highlighted by these two references is that linking is based on system alignment and compatibility, with equal unit values[100] that represent an amount of allowable emissions. All discussions of linking, including the first reference above, are based on this approach. As the review of the literature in the preceding subsection indicates, studies of linking date from when jurisdictions first began considering establishing domestic emissions trading, as a mechanism for mitigation, in the context of Kyoto Protocol commitments. Thus, conceptually, linking has developed in an environment where the traded units, the emission allowances, represented an amount of emission permitted under the particular scheme's cap, in effect, as an accounting unit for that scheme. Networking, on the other hand, is based on the converse, a system not of alignment but of heterogeneity, in which differences are respected and maintained, but valued according to agreed parameters. The traded units may differ in value, but importantly, they are valued for mitigation effectiveness, rather than representing an amount of emission allowed under a cap.

Networking has been described as 'heterogeneous linking' or 'soft linking'. This book argues such descriptions are misleading, as they incorrectly imply that networking is a form of linking. Linking, according to the second reference above, is a system in which the fungibility of traded units is based on a guiding principle of full system compatibility and equal unit value: what that could have gone on to say is that, not only is it based on 'equal unit value',

---

95   Ibid, 275.
96   Ibid.
97   Ibid, 276.
98   Ibid. See the discussion of a medium of exchange as a mechanism for effecting transactions at Chapter 7, following, which clarifies this point.
99   Ibid. See also fn.36 (Macinante).
100   Or in the case of restricted linking, relative proportions thereof.

but the units are all of the same kind, all being emission allowances (that is, rights to emit an amount of GHG equal to their value) often defined to equal one tonne CO_2-eq GHG. Linking has only ever been discussed in terms of links between ETSs, for good reason, because conceptually, linking has always been framed in terms of trading rights to emit.

In the alternative, this proposal posits that the metric of networking is units of mitigation achieved[101] by the particular mitigation action. Such a metric can be derived not just from ETSs, but as noted by Mehling, also from any other sorts of mitigation action. Thus, networking potentially can be across ETSs, carbon taxes, or even performance standards, provided there is an accepted methodology for determining the mitigation value of the outcomes of these mitigation actions. It is postulated that this difference affords networking significant advantages over linking as a way of connecting diverse and heterogeneous carbon markets to realize the potential benefits as have been discussed. The next chapter considers, inter alia, reasons why networking may afford a better approach than linking, to connecting markets, in elaborating the proposed market model.

---

[101] Also measurable in terms of tonnes of CO_2-eq GHG, but these are tonnes mitigated, removed, or abated, not tonnes that may be emitted.

PART III

The proposal

6. The networked market on distributed ledger technology – concept and theory

The market proposed here can be viewed as not a single market, but rather as a connection facilitating transactions between individual, separate markets, each of which will continue as an autonomous operation in its own jurisdiction, while participating in the network created by the connection. The proposal encompasses the digital infrastructure needed to provide the connection between these markets, as well as the legal and administrative structures that will operate, manage, and oversee the network. This chapter sets out the theory and concepts underpinning the proposed market.

The first section introduces the market proposed in this book in terms of its bifurcated nature; sets out the argument in favour of networking in preference to linking as a way to connect diverse carbon pricing schemes; and introduces the technology proposed to facilitate doing so. The second section examines that technology application in terms of specific characteristics, including in terms of the requirements of the Paris Agreement.[1] The rationale for the application of the technology is thereby derived.

The final section draws these threads together and leads into the following chapter, which elaborates what implementation of the proposal might entail in practical terms. Both this chapter and the next, to a significant extent, are dedicated to elaborating a new technology and a particular application of it, with the result that a certain level of descriptive material is included, while many of the sources referenced are from beyond the traditional, peer-reviewed academic literature. This range of sources provides a valuable contribution to building on what is currently an immature academic field. Importantly, it reflects also the inter-disciplinary elements of the subject, bringing together materials from a variety of fields.

---

[1]   FCCC/CP/2015/10/Add.1, 29 January 2016 <http://unfccc.int/resource/docs/2015/cop21/eng/10a01.pdf> accessed 13/03/17.

NETWORKED CARBON MARKETS ON DISTRIBUTED LEDGER TECHNOLOGY

1 .   The Two Elements of the Proposed Market

The proposed market is a network of carbon markets, on distributed ledger technology (DLT)[2] architecture. Thus, the proposal consists of two distinct elements, first, networking of carbon markets; and second, that networking being carried out using a specific type of digital information technology (IT) architecture, namely, a distributed ledger (or ledgers) (DL).

In addition to being comprised of these two distinct elements, the proposal can be viewed as proceeding down two independent, but interrelated, arms. The first of these can be seen as aiming to facilitate and stimulate an inter-jurisdictional market, so that it operates efficiently, encourages private sector engagement, promotes a stable carbon price, and fosters the effective application of carbon finance. This first arm is directed towards, and supports, the second arm, but can be seen also as providing a standalone outcome in its own right. The second arm of the proposal promotes the objectives of climate policy, evidenced by the terms of the Paris Agreement, including higher ambition, greater transparency, accuracy, accountability, and security of information sharing and management. This chapter will examine how characteristics of both elements, networking and DLT, contribute to and support both these arms of this proposal.

There is, at present, no trading network or market such as that which is proposed. Networking carbon markets is a concept introduced by the World Bank,[3] however, there are no existing examples, nor are there market networking models in other areas of application with which direct comparisons might be drawn. Even more so, while DLT use cases in the financial markets are being developed and will be considered, there are none that relate to a market of markets, proposed here. As such, the approach taken to analyzing the proposal is to consider the rationale for each element, in turn, independently of the other. For the networking element, reasons why such a connected trading arrangement between markets is desirable have been addressed in the preceding chapter. In the absence of an illustration of networking that might be examined, the reasons to network in order to achieve that connection, rather

---

[2]   DLT is sometimes referred to as 'blockchain', for reasons evident later in this section, although blockchain is just one implementation of the broader distributed ledger technology. This book will refer mostly to the broader concept, that is, DLT.

[3]   See generally: <http://www.worldbank.org/en/topic/climatechange/brief/globa lly-networked-carbon-markets> accessed 23/01/18.

than link, are drawn out in the following sub-section by considering issues that have arisen with linking and the extent to which networking might ameliorate or avoid them. For the DLT element, in relation to which, conversely, use cases are continuing to grow in a dynamic, developing environment, not only is it necessary to distinguish the use case proposed here from the expanding universe of such applications, but also to define what that use case is. This is set out in the third sub-section. The rationale for the application of DLT is then derived in the following section. The final section draws these two elements together. The details of the proposal are elaborated, in more tangible terms, in the following chapter.

2.   The Reasons to Network Rather than Link

(i)   Political issues

In order to determine that it is desirable to connect by linking, jurisdictions need to make a political decision, influenced by factors including perceived environmental stringency/credibility of the overall cap; perceived benefits, such as cost savings; impact on domestic action; distributional impacts; and loss of control.[4] With regards the last point regarding control, it would seem to be clearly preferable to avoid, as far as possible, compromising the sovereignty and autonomy of jurisdictions, as part of a process to engage them in a system of inter-jurisdictional cooperation. In this respect, networking has an advantage over linking in that, first, it requires less compromise of the domestic legal regime for trading, of the institutional structures, or of the independence of participating jurisdictions; and second, to the extent that it does involve compromise, any such accommodation by jurisdictions participating in the network will be required on the basis of equivalence: in other words, there would be a level playing field, where the same parameters would be applied equally to all.

Many political issues in relation to linking ETSs appear to flow from the potential impact it may have on jurisdictional sovereignty: for instance, the risk of design features from one jurisdiction's scheme extending to the scheme of the other, linked jurisdiction.[5] Networked jurisdictions' schemes, on the other

---

[4]   World Bank, Partnership for Market Readiness, *Lessons Learned from Linking Emissions Trading Systems: General Principles and Applications*, Technical Note 7, February 2014, 12. World Bank, Washington, DC. © World Bank.

[5]   Michael Mehling, 'Legal Frameworks for Linking National Emissions Trading Schemes' in C. Carlarne, K. Gray, and R. Tarasofsky (eds.), *Oxford Handbook of International Climate Change Law*, (Oxford University Press, 2016) 259 citing J. Jaffe and R. N. Stavins, *Linking Tradable Permit Schemes for Greenhouse Gas Emissions: Implications, and Challenges* (Geneva: International Emissions Trading Association,

hand, would remain separate, so the potential compromise of environmental objectives and control would not be an issue. Potential reduction in control over the domestic ETS, by its administrator, in a linked system would not arise in a networked arrangement and, similarly, the related issue of harmonization of ETS design elements would not arise, since the networked schemes remain separate and independent.

Linking results from an agreement negotiated by the governments of the respective jurisdictions seeking to link. Negotiations take time, sometimes a long time: in the case of Switzerland and the EU, for instance, seven years.[6] Inevitably, also, there will be imbalances between negotiating counterparties. An economically larger jurisdiction will, more than likely, have greater influence over the terms on which the parties link.[7] This is not to say that a smaller jurisdiction may be unwilling to accept that agreement, even though an unequal negotiating position may put it at a potential disadvantage.[8] Nevertheless, in such a negotiating process the smaller jurisdiction will be dependent, in some respects, on the goodwill of their larger counterparty.

The imbalance of negotiating positions should not arise in the case of networking. Rather, with networking the compromise of sovereignty – if it

---

2007); also, the challenges raised by linking are largely political in nature: Michael Mehling, 'Linking of Emissions Trading Schemes' in David Freestone and Charlotte Streck (eds.), *Legal Aspects of Carbon Trading: Kyoto, Copenhagen and Beyond* (Oxford University Press, 2009).

6   EC Climate Action announcement 23/11/17 that EU and Switzerland had signed an agreement to link their emissions trading schemes, noting that negotiations opened in 2010: <https://ec.europa.eu/clima/news/eu-and-switzerland-sign-agreement> accessed 18/12/17.

7   The EUETS is cited as an example of a unilateral approach under which other carbon markets have to adapt to its architecture, although the California-Quebec negotiation is, on the contrary, collaborative: Dmitry Fedosov, 'Linking Carbon Markets: Development and Implications' [2016] *CCLR* 202. However, both California and Quebec are part of the initial collaboration, the Western Climate Initiative and their schemes were very similar to begin with: Christiane Beuermann et al., 'Considering the Effects of Linking Emissions Trading Schemes, A Manual on Bilateral Linking of ETS', May 2017, German Emissions Trading Authority (DEHSt) on behalf of German Environment, 13 <https://www.dehst.de/SharedDocs/downloads/EN/emissions-trad ing/Linking_manual.pdf?__blob=publicationFile&v=3> accessed 17/07/17. Note also that California and Quebec staff conducted line-by-line comparisons of the respective program regulations in order to harmonize them in every respect to ensure environmental integrity and compatibility: fn.4 (World Bank, PMR, 2014) 15.

8   The experience to date has been that when linking with the EUETS, the other scheme needs to align itself with the EUETS; Switzerland revised its ETS in December 2011, to increase compatibility with the EUETS, see: Angelica P. Rutherford, 'Linking Emissions Trading Schemes: Lessons from the EU-Swiss ETSs' [2014] *CCLR* 282.

could be called that – would come in the form of acceptance of the parameters by which a jurisdiction's mitigation actions are valued (to give the mitigation value (MV) of the jurisdiction's mitigation outcomes).[9] These parameters would apply on the same basis to all jurisdictions that agree to participate in the network. Hence, the compromise would apply equally to all participating jurisdictions, rather than differentially depending on the relative economic size of counterparties to the particular bilateral or multilateral linking arrangement.

In linked systems, convergence of prices may have distributional impacts on participants and other stakeholders, resulting in substantial capital flows that may affect political support.[10] While there is no reason to expect there would not be distributional impacts also in any networked system, it is envisaged that networking arrangements would afford governments greater flexibility to set the terms for participation by the entities they authorize to trade in the networked market. Networking would also incorporate greater flexibility for a jurisdiction to opt out altogether were it to determine that trading flows are no longer favourable to its domestic policy objectives.

A further consideration is that linking arrangements can default to the lowest mitigation standard of those jurisdictions participating, thereby affecting jurisdictions whose policies target higher ambition.[11] In a networking arrangement, the aim of market design is to correlate MV with price, so that the market incentivizes continued improvement, in conformity with the Paris Agreement objectives seeking higher ambition.[12] Thus, the aim would be for the market to operate so as to encourage a race to the top, not the bottom. It is appreciated

---

[9]   See: Justin D. Macinante 'Operationalizing Cooperative Approaches Under the Paris Agreement by Valuing Mitigation Outcomes' [2018] *CCLR* 258: discussed in Chapter 7 following.

[10]   Fn.5 (Mehling 2016) 259 citing: R. Baron and C. Philibert, 'Act Locally, Trade Globally Emissions Trading for Climate Policy', © OECD/IEA, 2005 <https://www.iea.org/publications/freepublications/publication/act_locally.pdf> accessed 14/05/17; also see: Matthew Ranson and Robert N. Stavins, 'Linkage of Greenhouse Gas Emissions Trading Systems: Learning from Experience' Discussion Paper ES 2013-2. Cambridge, Mass.: Harvard Project on Climate Agreements, November 2013, 9; the nature of impacts will also be a function of the elasticity of demand in certain markets (that is, whether the additional costs can be passed through to consumers) and the extent to which regulated entities are competing in international markets not covered by emission mitigation restrictions, see: Mirabelle Muûls et al., 'Evaluating the EU Emissions Trading System: Take it or leave it? An assessment of the data after ten years', October 2016, Imperial College London, Grantham Institute, Briefing Paper No. 21 <https://www.imperial.ac.uk/media/imperial-college/grantham-institute/public/publications/briefing-papers/Evaluating-the-EU-emissions-trading-system_Grantham-BP-21_web.pdf> accessed 16/03/17.

[11]   Fn.4 (World Bank, PMR, 2014) 9.

[12]   For example: Article 2 and Article 4, paragraph 3, Paris Agreement.

that assessments valuing mitigation outcomes may cause consternation for some governments. Nevertheless, the mitigation outcomes of all participating jurisdictions would be assessed independently, according to the same objective, technical criteria. Participating jurisdictions would have assessments made on the same basis, such that there would be equivalence of treatment. Additionally, the feedback from and the transparency of the MV assessment process (discussed in the following chapter) should enhance jurisdictions' information and knowledge bases, again, facilitating continuous improvement.

As mentioned earlier, linking has been described as being politically complex and this has been suggested as a reason for so few links occurring to date,[13] although there seems to be a divergence of views on the extent to which linking has actually been occurring.[14] Establishing an operational system for trading mitigation outcomes, based on a network between jurisdictions, could also well involve elements of political complexity. All the same, it is posited that many of the issues and obstacles, such as those outlined above, that complicate, slow, or deter attempts to link jurisdictions are not present, or not present to the same extent, in the case of networking. From a political perspective, networking offers a more flexible way to achieve international transfers of mitigation outcomes.

(ii)   Legal issues

It has been pointed out that political motives are not all that need to be considered in relation to linking: ETSs operate in complex frameworks of rules, principles, and procedures under domestic law[15] and these factors will be relevant when units are traded across jurisdictions. Just as with linking of ETSs, networking would require agreement between participant jurisdictions. In the case of networking, however, the nature of the agreement is fundamentally different. Rather than being between two (or more) individual jurisdictions, each of which is seeking to construct the arrangement on its own terms in order to reduce the degree to which it must compromise its existing system, a networking agreement would be between the jurisdiction seeking to join and the network, that is, the platform on which trading takes place.

---

[13]   Fn.5 (Mehling, 2016) 258.
[14]   Fn.4 (World Bank, PMR, 2014), according to which, bilateral linking was rare to date of that publication but other authors are more bullish about links up to the date of the Paris Agreement: see Michael A. Mehling, Gilbert E. Metcalf, and Robert N. Stavins, 'Linking Heterogeneous Climate Policies (Consistent with the Paris Agreement)' Discussion Paper ES 2017-6. Cambridge, Mass.: Harvard Project on Climate Agreements, October 2017. Mehling seems to have changed from previous position (Fn.5 (Mehling 2016)).
[15]   Fn.5 (Mehling, 2009) 116.

Under the networking arrangement there would not be a need for agreement as to legal alignment of parties' ETSs to ensure the respective units are fungible; there would not be a need for joint registries, nor would there be a need for joint auctioning, or similarly coordinated issuance arrangements. Under networking, the jurisdictions' ETSs would remain independent of each other.[16] Thus, there would be no need to harmonize the regulatory systems,[17] institutions, administration, or procedures, simplifying the process for jurisdictions to decide whether to participate, or not.

This approach relies on the jurisdictions that wish to participate in the network accepting the rules, infrastructural arrangements, and other measures – such as the mechanism and parameters for determining the value of participating jurisdictions' mitigation actions (the MV) – and adhering to those rules and other requirements. As proposed here, the agreement required of a prospective networking participant would involve, first, acceptance of the same terms on which all other jurisdictions agree to participate; and second, that jurisdiction signifying any limits or conditions it wishes to impose on transactions entered by the legal entities it authorizes to trade on the network. The decision whether to join – at least in so far as the terms and conditions of participation – should be relatively straightforward: either accept and join, or reject and not join. As noted above, the agreement is not between jurisdictions, as such, but rather between the joining jurisdiction and the network (that is, the collective of jurisdictions that have already agreed to the common rules). Further, as matters such as ETS alignment, registries, and issuance do not need to be negotiated, the relative bargaining position of jurisdictions, as a legal issue in negotiations, is rendered nugatory.

A further legal consideration relates to the nature of what is being traded, that is, the carbon units. One author has argued, for instance, that while the nature of carbon units makes it difficult to classify them under traditional categories of financial instruments, nevertheless, it is likely that they constitute objects regulated by the Annex on Financial Services to the General Agreement on Trade in Services (GATS) under the WTO Agreement.[18] This finding is

---

[16] This is subject to the qualification that under the proposal, the ledger (registry) is distributed, such that all participating jurisdictions may hold a copy of the ledger for all transactions across the entire network: see following sections.

[17] For instance, Article 4 of the California-Quebec linking agreement provides specifically for regulatory harmonization: see Agreement between California Air Resources Board and the Government of Quebec, Concerning the Harmonisation and Integration of Cap-And-Trade Programs for Reducing Greenhouse Gas Emissions, 27 September 2013 <https://www.arb.ca.gov/cc/capandtrade/linkage/ca_quebec_linking _agreement_english.pdf> accessed 06/03/18.

[18] James Munro, 'Trade in Carbon Units as a Financial Service under International Trade Law: Recent Developments, Future Challenges' [2014] *CCLR* 106, 113.

applied to conclude that carbon markets are subject to international trade rules, which could thereby lead to emission trading schemes that only accept their own units, or perhaps also units from other linked schemes, being impugned.[19] However, this does not seem to address the issue of whether surrender of the units against compliance obligations, by an entity regulated domestically under the scheme (which is, after all, the core element of a scheme where the restriction on units becomes relevant) comes within the concept of trade under the GATS, nor the question of the equivalence of those units.

All the same, even though it does not need to be answered for the purposes here, the argument is interesting for two reasons. First, in the context of multinational, or global corporations carrying on all sorts of trading and other business undertakings in multiple jurisdictions, thus being exposed to the growing number of emissions trading schemes spawning across those juris-dictions, it may only be a matter of time before an such issue such is tested. Second, and pursuant to the preceding point, it would seem logical, therefore, for climate policymakers to embrace a trading mechanism that could both (a) facilitate emission unit trading across jurisdictional boundaries, doing so within a climate policy framework that accounts for the differences between jurisdictions, but (b) in the converse, could supply a methodology (that is, MV assessment) to show that discrimination is objectively necessary and reasona-ble in terms of climate mitigation,[20] thereby justifying, as an alternative, reli-ance on an exception under the GATS, when necessary. Networking provides such a trading mechanism.

(iii)   Practical issues

It follows from the political and legal considerations that in terms of practical application, networking should be administratively more feasible than linking. As proposed here, networking would not require the transfer of units from one registry to another, thereby avoiding the legal and administrative complexity that can arise in proposals for linking arrangements. There would still need to be the physical (electronic) infrastructure to give effect to transactions, but unlike approaches to linking, networking would not require equivalence of the assets – emission allowances – in the connecting ETSs in order to achieve fungibility.

In a networked system, the units may not even need to be the same type of asset, or primarily measured in the same terms (for example, the asset in one scheme might be measured as an absolute value, in the other as a performance standard), provided an assessment can be made of the respective mitigation

---

[19] Ibid.
[20] Ibid, 114.

values. Thus, subject to agreement being reached on (or at least there being acceptance of) the parameters and methodology for comparative assessments, the actual transaction process should be simpler, and more transparent.

Two further points follow from the greater simplicity of the transaction process in a networking arrangement: first, since there is no need to move units from the registry of one ETS to the registry of another, accounting and record keeping in a networked system would be less complicated. Instead, the units the MV of which is to be transferred would be cancelled in their domestic registry (an illustration of how transactions might proceed is in the following chapter); and second, once applicable parameters and methodology for comparative assessments have been agreed, networking would not be restricted to ETSs, but could include other mitigation actions, provided their outcomes were capable of MV assessment. As such, networking offers potential scope for a much larger, more flexible market than could occur under linking. It is considered this would also be more effective in re-engaging the private sector.

Connecting ETSs, whether by linking or networking, necessarily involves reconciling the differences between schemes. The integral point of difference is the extent of mitigation brought about by the respective schemes. By assessing MV, the networking approach separates this climate element from elements of a more administrative or mechanistic nature, whereas linking requires the harmonization of these elements as part of the process to reconcile climate (mitigation) element differences.[21]

(iv)   Flexibility (opting in and out)
It follows also from the preceding points, that because there is no need for legal, institutional, or administrative integration of systems, it is more flexible for jurisdictions to join or leave the networked market. The network, in this sense, might be viewed as a facility of which any jurisdiction might avail itself, so long as it sees there is an advantage for participants in its domestic market, and from which it might remove itself when it perceives that advantage no longer continues. While there would be a need for institutional and regulatory frameworks for the network itself and these would require time and resources to establish, their existence and operation should not inhibit the flexibility of jurisdictions seeking to join or leave the network, but rather facilitate it.

Two complementary consequences flow from this structure: first, the network could continue to operate unaffected when an individual jurisdiction elects to leave it; and second, a jurisdiction that wishes to opt out of the

---

[21]   In this respect, the line-by-line comparisons of the respective program regulations by California and Quebec staff spring to mind: see fn.4 (World Bank, PMR, 2014) 15.

network could do so seamlessly, not only without impacting ongoing network operation, but also without affecting operation of its own ETS. For individual jurisdictions this would mean less of an administrative burden, less cost, and the ability to give effect to decisions relatively expeditiously – certainly much more quickly than the time it would take to negotiate a linking agreement, or the severing of one.

3.   Distributed Ledger Technology

(i)   Introduction

This sub-section introduces the second element of the proposal, the specific type of IT architecture in which the network of carbon markets might operate. In this respect, the question might be posed why is it necessary, or even desirable, to specify as part of this proposal for a market to achieve climate objectives, the IT platform architecture on which it is to operate? The short answer is that, like any other financial market, it is being driven by technological change. This and the following sub-sections on use cases, terminology, definitions, and the use case of the proposal, expand on that answer.

The proposal for networking of carbon markets across jurisdictions necessarily implies that there must be some form of infrastructure (which, it is assumed, would necessarily need to be electronic) in place to allow such a market to operate by transactional communications taking place between participants, even if this were just some basic form of IT communication, say, by email across the internet. What is proposed, however, is the inter-jurisdictional trade in carbon assets and, as outlined in an earlier chapter, increasingly these are being defined legislatively as financial instruments.[22] As such, the networked market is proposed to be a financial market (albeit, one constructed for the purpose of achieving environmental objectives), implying certain basic essential requirements for its transactional infrastructure, such as security, capacity, and reliability.[23] As with any financial market, this infrastructure might be expected to facilitate accountability, auditability, certainty, and accuracy of the transactions it processes, as well as regulatory supervision, the facility to ensure financial and legal risk management can be addressed, and that the system's capacity is as time and cost efficient, as possible.

---

[22]   See, for instance: Markets in Financial Instruments Directive II (MiFID II): Directive 2014/65/EU of the European Parliament and of the Council of 15 May 2014 on markets in financial instruments and amending Directive 2002/92/EC and Directive 2011/61/EU, OJ L 173, 12.06.2014, 394–496.

[23]   As to characteristics of a financial market generally, see for instance: Shelagh Heffernan, 'A Characteristics Definition of Financial Markets' (1990) 14(2–3) *Journal of Banking and Finance* 583.

The context in which this proposal is made is one of global recognition that technological developments are occurring that will fundamentally change how financial services are provided, how markets, business, and governments operate.[24] These developments are occurring in many fields of application, at such a rate of change that it is difficult to present an overview with more than a pretence of completeness, or one that might remain so for any length of time. All the same, they include developments in subject areas such as Big Data;[25] Internet of Things;[26] the platform economy;[27] and in so-called emerging transformative technologies that include biometrics; cloud computing; cognitive computing; distributed ledger technology (DLT), or blockchain; machine learning, or predictive analytics; quantum computing; and robotics.[28]

To a degree there is overlap across these technological areas; nevertheless, the focus of this proposal is solely on DLT and blockchain (an introductory description is set out in the first chapter). This technology alone has been described as portending 'a new digital revolution',[29] coming after 20 years of scientific research that produced advances in the fields of cryptography and decentralized computer networks.[30] Such exorbitant claims may not be as outlandish as sober assessment would otherwise suggest, given the level

---

[24]   Mark Walport, Chief Scientific Adviser to HM Government, 'Distributed Ledger Technology: beyond block chain' A report by the UK Government Chief Scientific Adviser, UK Government Office for Science, GS 16-1, published 19/01/16 <https://www.gov.uk/government/publications/distributed-ledger-technology-blackett-review> accessed 30/09/16; Carlota Perez, 'Technological Revolutions and Techno-Economic Paradigms' (2010) 34(1) *Cambridge Journal of Economics* 185, 197 Table 3: this is the fifth technological revolution: the Age of Information and Telecommunications, see innovation principles.

[25]   Gartner   (2012)   <https://www.gartner.com/it-glossary/big-data>   accessed 8/01/18.

[26]   IEEE, 'Towards a definition of the Internet of Things (IoT)' Revision 1 published 27 May 2015 (IEEE) <https://iot.ieee.org/definition.html> accessed 8/01/18.

[27]   For discussion of definitions and approaches to regulation, see: Michèle Finck, 'Digital Co-Regulation: Designing a Supranational Legal Framework for the Platform Economy' (2018) 43(1) *European Law Review* 47.

[28]   For how financial services industry transformation has spun off technology innovation over the last 50 years, see: World Economic Forum, 'The future of financial infrastructure: An ambitious look at how blockchain can reshape financial services' (WEF, New York USA, August 2016), 20 <www.wef.org> accessed 02/11/16.

[29]   Aaron Wright and Primavera De Filippi, 'Decentralized Blockchain Technology and the Rise of Lex Cryptographia' Background Paper (Mar 12, 2015) Internet Governance Forum, UN-Department of Economic and Social Affairs, Workshops Descriptions and Reports, IGF 2015 Workshop No.239 Bitcoin, Blockchain and Beyond: FLASH HELP! 2 <http://www.intgovforum.org/cms/workshops/list-of-published-workshop-proposals> accessed 3/11/16.

[30]   Ibid.

of attention and related research being applied by intergovernmental bodies, governments and public institutions,[31] global business bodies,[32] the financial sector,[33] lawyers and consultants,[34] and market regulators.[35]

---

[31]   For example, to mention a few: Bank of International Settlements: Morten Bech and Rodney Garratt, 'Central Bank Cryptocurrencies', September 2017, *BIS Quarterly Review*, 55–70 <https://www.bis.org/publ/qtrpdf/r_qt1709.pdf> accessed 24/01/18; European Commission: Communication from the Commission to the European Parliament, the Council, the European Economic and Social Committee and the Committee of the Regions, Online Platforms and the Digital Single Market Opportunities and Challenges for Europe, COM/2016/0288 final; Robleh Ali, John Barrdear, and Roger Clews, 'Innovations in payment technologies and the emergence of digital currencies', Bank of England Quarterly Bulletin, 2014 Q3 <http://www.bankofengland.co.uk/publications/Documents/quarterlybulletin/2014/qb14q3d igitalcurrenciesbitcoin1.pdf> accessed 12/01/17; John Barrdear and Michael Kumhof, Bank of England, Staff Working Paper No.605, 'The macroeconomics of central bank issued digital currencies', July 2016, (Bank of England 2016) <http://www .bankofengland.co.uk/research/Documents/workingpapers/2016/swp605.pdf> accessed 12/01/17; A. Blundell-Wignall, 'The Bitcoin Question: Currency versus Trust-less Transfer Technology', OECD Working Papers on Finance, Insurance and Private Pensions, 2014, No.37, OECD Publishing <http://dx.doi.org/10.1787/ 5jz2pwjd9t20-en> accessed 27/10/16.

[32]   Fn.28 (World Economic Forum).

[33]   For example, R3 is a consortium with over 80 banks, clearing houses, exchanges, market infrastructure providers, asset managers, central banks, conduct regulators, trade associations, professional services firms and technology companies developing commercial applications of distributed ledger technology for the financial services industry: <http://www.r3cev.com/blog/2016/4/4/introducing-r3-corda-a-distributed -ledger-designed-for-financial-services> accessed 12/03/18.

[34]   For example: Sigrid Seibold and George Samman, 'Consensus: Immutable agreement for the Internet of value', KPMG, (2016), <https://assets.kpmg.com/content/dam/ kpmg/pdf/2016/06/kpmg-blockchain-consensus-mechanism.pdf > accessed 05/02/18; Allens Lawyers, 'Blockchain Reaction Understanding the opportunities and navigating the legal frameworks of distributed ledger technology and blockchain', report <http://www.the-blockchain.com/docs/blockchainreport-%20legal%20frameworks %20of%20distributed%20ledger.pdf> accessed 02/11/16; also see, in general: <www .lexology.com/blockchain>.

[35]   For instance: European Securities and Markets Authority (ESMA), 'The Distributed Ledger Technology Applied to Securities Markets' Discussion Paper, 2 June 2016, ESMA/2016/773; Swiss Financial Market Supervisory Authority FINMA, Guidelines for enquiries regarding the regulatory framework for initial coin offerings (ICOs), Published 16 February 2018, <https://www.finma.ch/en/news/2018/02/ 20180216-mm-ico-wegleitung/> accessed 23/02/18; Financial Conduct Authority, Discussion Paper on Distributed Ledger Technology, DP17/3, April 2017 <https:// www.fca.org.uk/publication/discussion/dp17-03.pdf> accessed 19/04/17; Financial Conduct Authority UK, Distributed Ledger Technology Feedback Statement on Discussion Paper 17/03, Feedback Statement FS17/4, December 2017 <https://www .fca.org.uk/publication/feedback/fs17-04.pdf> accessed 22/12/17; European Banking

While much of this research attention is applied to the opportunities and potential benefits the technology offers,[36] some applications, such as crypto-currencies and their uses, as well as aspects that test the boundaries of current regulations, such as initial coin offerings (ICOs), are increasingly the focus of lawmakers' and regulators' attention.[37] The applications of DLT for business, financial, and government services, while growing rapidly, are still nascent, yet already there has been consideration given in the literature to the regulation of DLT[38] and this is increasing as new use cases are assessed and implemented. Nevertheless, historically it remains largely limited to and focused on specific applications of the technology such as cryptocurrencies.[39]

Finally, in this respect, it is noted there have been issues, previously, with security of carbon market transactions and the existing carbon market IT.[40] The expectation is that ongoing technological developments can help ensure episodes such as hacking of registry accounts are far less likely to recur, if not

---

Authority, Opinion on the EU Commission's proposal to bring Virtual Currencies into the scope of the Directive (EU) 2015/849, EBA-Op-2016-07, 11 August 2016; ASTRI Whitepaper On Distributed Ledger Technology (11 November 2016) Commissioned by Hong Kong Monetary Authority <http://www.hkma.gov.hk/media/eng/doc/key-functions/finanical-infrastructure/Whitepaper_On_Distributed_Ledger_Technology.pdf> accessed 12/1/17.

[36] See, for example: Michèle Finck, 'Blockchains: Regulating the Unknown' (2018) 19(4) *German Law Journal* 665; Julie Maupin, 'Mapping the Global Legal Landscape of Blockchain and Other Distributed Ledger Technologies', Centre for International Governance Innovation, CIGI Papers No.149, October 2017. <https://www.cigionline.org/sites/default/files/documents/Paper%20no.149.pdf> accessed 24/01/18; Benno Ferrarini, Julie Maupin, and Marthe Hinojales 'Distributed Ledger Technologies for Developing Asia', ADB Economics Working Paper Series, No. 533 | December 2017. <https://www.adb.org/sites/default/files/publication/388861/ewp-533.pdf> accessed 24/01/18; Angela Walch, 'The Path of the Blockchain Lexicon (and the Law)' (2017) 36(2) *Review of Banking & Financial Law* 713.

[37] Ibid (Finck). See also Chapter 8, following.

[38] See also Chapter 8.

[39] In the US, for instance, see: Financial Crimes Enforcement Network (FinCEN), US Department of Treasury <http://www.fincen.gov>; US Commodities Futures Trading Commission Act 1974 (7 U.S.C. §§ 1 et seq) <http://www.cftc.org>; New York Department of Financial Services (NYDFS) 'BitLicence' N.Y. Comp. Codes R. & Regs. Tit.23, § 200 (2015) <http://www.dfs.ny.gov/>; Securities and Exchange Commission, Securities Exchange Act of 1934 Release No. 81207/July 25, 2017 Report of Investigation Pursuant to Section 21(a) of the Securities Exchange Act of 1934: The DAO <https://www.sec.gov/litigation/investreport/34-81207.pdf> accessed 21/08/17.

[40] European Court of Auditors Special Report 'The integrity and implementation of the EU ETS' 2015, 29–41. <http://www.eca.europa.eu/Lists/ECADocuments/SR15_06/SR15_06_EN.pdf> accessed 23/06/17.

impossible.[41] Additionally, this technological development purports to hold out the promise of better addressing some of the core elements of climate policy incorporated in the Paris Agreement, such as greater transparency, accountability, traceability, and security. The extent to which DLT could better address these elements, than existing IT infrastructure does, is considered in the section on specific characteristics of the technology.

(ii)   Use cases, especially in financial markets

A World Economic Forum (WEF) report in 2016 found that, while there was significant awareness and interest in DLT, hurdles to large-scale implementation (in terms of financial infrastructure), such as an uncertain and unharmonized regulatory environment, nascent collective standardization efforts, and an absence of formal legal frameworks, remained.[42] Some of the potential areas of application of DLT that have been identified include in trade finance, through operational simplification; in compliance automation, improving regulatory efficiency; in global payment systems, by reducing settlement times; and in asset rehypothecation, thereby enhancing liquidity.[43] Other broader, potential applications being tested or implemented include in relation to record keeping, such as patient health records, or land property titles; legal inheritance; source traceability for supply chains, including diamonds, or gold production; or other proof of ownership.[44] Some intergovernmental bodies,

---

[41]   There are structural issues as well. For instance, in 2012, the centralized Union registry replaced all national registries in the EUETS: Commission Regulation (EU) No 389/2013 of 2 May 2013 establishing a Union Registry pursuant to Directive 2003/87/EC of the European Parliament and of the Council, Decisions No 280/2004/EC and No 406/2009/EC of the European Parliament and of the Council and repealing Commission Regulations (EU) No 920/2010 and No 1193/2011 Text with EEA relevance, OJ L 122, 3.5.2013, 1–59; although, the European Court of Auditors report noted that even though the EC operates the EU registry, it has no powers to monitor and supervise transactions: fn.40 (ECA) supra.

[42]   Fn.28 (World Economic Forum): reported that at that time, more than 24 countries were investing in DLT, over 90 corporations had joined DLT consortia, 80 per cent of banks were predicted to initiate DLT projects by 2017 and over the preceding three years, more than 2500 patents had been filed and over US$1.4 billion invested; see also fn.34 (Allens Lawyers); Stuart Davis and Julian Cunningham-Day, Linklaters LLP, 'Blockchain – recalibrating the market infrastructure', Going Digital Quarterly Breakfast Briefing, 14 October 2016, presentation Powerpoint slide deck.

[43]   Ibid (World Economic Forum) 21. For example, also reported that the Hong Kong Monetary Authority was leading a project with 21 banks to provide a blockchain-based trade finance platform to enhance efficiency, reduce transaction costs: Financial Times, 16 July 2018, 18.

[44]   Fn.34 (Seibold and Samman/KPMG); fn.36 (Maupin/CIGI 2017); International Monetary Fund, 'Virtual Currencies and Beyond: Initial Considerations' January 2016, Staff Discussion Note SDN/16/03.

national and provincial governments have instigated projects to provide services based on DLT.[45] Other application areas that have been reported include decentralized power generation sharing, music streaming royalty payments, and voting in elections.[46]

In terms of potential areas of impact of DLT on financial markets, operational simplification, regulatory efficiency improvement, counterparty risk reduction, clearing and settlement time reduction, liquidity and capital improvement, and fraud minimization have been identified as value drivers.[47] The claimed 'transformative characteristics' of distributed infrastructure include immutability, which for financial market participants might remove the need for reconciliations;[48] transparency, thereby removing market information asymmetries and increasing regulator/regulated party cooperation; and autonomy, disintermediating centralized parties whose roles in bringing trust and reducing counterparty risk will be obviated.[49] Possible benefits of DLT applied, for instance, to the securities market include speeding up clearing and settlement by reducing the number of intermediaries involved; facilitating recording of ownership and safekeeping of assets; facilitating collection, consolidation, and sharing of data for reporting, risk management, and supervisory purposes; reducing counterparty risk by shortening the transaction settlement cycle; improving the efficient management of collateral; continuous availability; greater security and resilience against attack; and cost reduction.[50] Other possible financial services applications relate to global payments, trade finance, corporate proxy voting, insurance claims processing, syndicated loans, and contingent convertible bond issuances.[51]

---

[45]   Fn.36 (Walch) 718, n.13; also fn.24 (Walport) chapter 6; fn.31 (Ali et al./ Bank of England); Reuters report on UN using blockchain to avoid fraud in aid shipments <https://www.reuters.com/article/us-un-refugees-blockchain/u-n-glimpses -into-blockchain-future-with-eye-scan-payments-for-refugees-idUSKBN19C0BB> accessed 28/01/18; fn.36 (Ferrarini et al./ADB 2017), give examples of digital identity, trade finance, project aid monitoring and results-based disbursements, smart energy, and sustainable supply chain management.

[46]   Fn.36 (Finck) 671–4; Marc Pilkington, 'Blockchain Technology: Principles and Applications' in F. Xavier Olleros and Majlinda Zhegu (eds.), *Research Handbook on Digital Transformations* (Edward Elgar, 2016); The Guardian, article on broader applications of blockchain <https://www.theguardian.com/technology/2018/jan/28/ blockchain-so-much-bigger-than-bitcoin> accessed 28/01/18.

[47]   Fn.28 (World Economic Forum) 19 et seq.

[48]   Note that this characteristic is explored in more detail in the following section.

[49]   Fn.28 (World Economic Forum) 24; also fn.34 (Allens Lawyers); fn.42 (Davis et al./Linklaters).

[50]   Fn.35 (ESMA 2016) 9–13.

[51]   Fn.28 (World Economic Forum) 46–127, setting out 'deep dive analyses of these and other use cases'.

Notwithstanding the overwhelmingly positive sentiment that surrounds the applications and benefits to be expected of DLT, it is important not to be swept up by the hype of the 'thought leaders'.[52] The general perception, in this environment, is that DLT could make networking of carbon markets both feasible and effective, by enabling traceability of the provenance of assets, or their attributes such as mitigation value; by the security dimension it brings; and by the permanence of records it can afford, thereby facilitating accounting and auditability. Thus, it would be promoting the objectives of climate policy, evidenced by the terms of the Paris Agreement, while also facilitating and stimulating an inter-jurisdictional market, so that it operates efficiently, encourages private sector engagement, promotes a stable carbon price and fosters the effective application of carbon finance. These perceptions are examined below.

(iii)   DLT terminology

The dynamic state of DLT development and the range of fields in which it might be applied introduce issues of terminology and meaning.[53] For example, the expressions 'DLT' and 'blockchain' are frequently used interchangeably, both in academic and general literature. Even use of 'distributed' can cause the misperception that because a ledger is distributed, there is no overall controlling entity, whereas this is a question of design.[54] Confusion of meaning over the terms used is a risk not only for academics, researchers, and business entities designing and building applications in the various different fields, but more especially so for policymakers and regulators overseeing such developments and determining the extent to which their intervention in the use cases is warranted and how that intervention should be carried out.[55]

The technology is populated with particular nomenclature such as 'permissioned' and 'permissionless', 'smart contracts', 'miners' and 'mining', 'tokens', 'cryptocurrencies', 'initial coin offerings' and with acronyms, such as, just in relation to different types of cryptography and security, PKI, HASH, SHA-256, zk-SNARK, HE, ECC, ECDSA, SGX.[56] For some expressions, there will be other parallel expressions (for example, public and private, for

---

[52]   Fn.36 (Walch) 740, n.108.
[53]   Fn.24 (Walport) 7, refers to 'the bewildering array of terminology' as a difficulty in communication.
[54]   Ibid.
[55]   Fn.36 (Walch) 728 et seq.
[56]   Mark Simpson and Steven Wang, 'Bitcoin, Crypto Assets and Blockchain', RBS Emerging Technology presentation at Edinburgh University, 8 February 2018 <https:// uoe.sharepoint.com/:p:/r/sites/sw-dev-group/_layouts/15/Doc.aspx?sourcedoc= %7B55954a3c-3de6-4ae0-a3ae-2788d521c4d9%7D&action=edit> accessed 13/02/18.

permissionless and permissioned), which may have identical meanings, or slightly nuanced differences of meaning.[57]

Of perhaps greater concern is the way in which fundamental descriptive characteristics of the technology may be understood, particularly when they are used so broadly and repetitively that they enter the technological/DLT vernacular without scrutiny or detailed consideration. Walch cites the example of 'immutable', as used to describe the ledger created by blockchain technology, in this respect.[58] In view of the integral importance of it as a characteristic of DLT, since other claimed beneficial characteristics of the technology, such as traceability, accountability, and auditability follow from it, immutability is considered in more detail in the next section, along with other such characteristics.

(iv)   DLT definitions

In the shifting sands of terminology flagged above, formal definitions will not necessarily be universally agreed and, even so, may be superseded relatively quickly.[59] All the same, it is necessary to clearly explain what is meant by the terms and expressions, as employed here, in this particular context.[60]

The infrastructure on which it is proposed to provide networking of carbon markets is, at its most elementary, a series of computers, or nodes, connected with each other in a network, for instance, via the internet. In this sense, it is no different from other such structures that exist, for example, the connections of computers of legal entities trading in the EUETS or other markets. The fundamental difference introduced by DLT is that the ledger, or registry – the record of unit holdings of participating entities resulting from the transactions between them – is no longer held only by a trusted, centrally positioned entity (comparable, for instance, to the International Transaction Log (ITL) under the Kyoto Protocol, although the ITL role is also more limited) through which all transactions must be routed in order to be approved, recorded, and that record maintained. Rather, the ledger may be held in full and kept up-to-date on all nodes, that is, on each participating entity's computer (or alternatively, just on a certain number thereof). Thus, the ledger is distributed. Another description is as a shared ledger, which has been applied particularly in the context of industry-based (e.g., financial sector) applications.[61]

---

[57]   Fn.36 (Walch) 719–28, has examined this issue in considerable detail, highlighting the particular problems this generates for regulators.

[58]   Ibid, 735–45.

[59]   See, for instance, fn.36 (Walch) 730 in relation to New York's 'BitLicence'.

[60]   Fn.24 (Walport) 17–19.

[61]   Ibid.

DLT is considered broadly as consisting of three elements, being the combination of a distributed ledger, with public/private key encryption, and a decentralized infrastructure.[62] It has been described also as 'a distributed, shared, encrypted-database that serves as an irreversible and incorruptible public repository of information', enabling 'unrelated people to reach consensus on the occurrence of a particular transaction or event without need for a controlling authority'.[63] Another description of DLT is as 'a protocol for building a replicated and shared ledger system', collectively maintained by the participants in that system or network, rather than by one central party.[64]

DLT is not a huge technological leap, but rather an incremental improvement,[65] one source even noting the existence of ledgers over thousands of years.[66] In DLT, the ledger can (but need not necessarily) be organized as a chain of blocks of information, each block containing a collection of transactions – new transactions being collected to form a new block that is time-stamped when added to the ledger.[67] Each block, thus, contains one or more new transactions and the adding of blocks to the chain (hence this implementation is referred to as the 'blockchain') means the ledger grows cumulatively.[68]

Blockchain is one implementation of a distributed ledger. Records can also just be stored one after the other, on a distributed ledger, in a continuous manner (but not in blocks), being added after the participants reach consensus.[69] There is also a newer type of DLT that uses Directed Acyclic Graphs (DAGs) that transmit and confirm transactions in an asynchronous, as opposed to chained way.[70] However, it is not necessary for these purposes to catalogue and examine every such form, except to the extent it impacts on the beneficial characteristics of the DL and, ultimately, the regulatory framework. It is simply noted that different technical mechanisms exist for adding to the ledger.

---

[62]   For example, see: fn.35 (ESMA 2016) section 2.1; also fn.29 (Wright and De Filippi) 4, 5.
[63]   Fn.29 (Wright and De Filippi) 2.
[64]   Fn.35 (ASTRI).
[65]   Fn.29 (Wright and De Filippi) 5, note 15. These authors trace the historical development of the individual elements back to the late 1970s. See also fn.28 (World Economic Forum).
[66]   Fn.35 (ASTRI).
[67]   Ibid.
[68]   Ibid.
[69]   Fn.24 (Walport) 18.
[70]   Fn.36 (Ferrarini et al./ADB 2017) 5.

As DLT covers a wide range of potential functionality, it is useful to identify key features that define a DLT system.[71] These are:

- first, a decentralized, distributed infrastructure, meaning the system is composed of multiple entities or nodes, each (or at least a number thereof) holding a copy of the full ledger, obviating the role of the central ledger holder;
- second, participants using public/private key encryption to interact with transactions in the system, obviating the role of a trusted central counter-party to intermediate transactions;
- third, a mechanism by which the nodes reach consensus on the valid entries to add to the ledger; and
- fourth, immutability, meaning that the ledger is accumulative, so that once entries are added to the ledger, (theoretically, at least) they cannot be changed or removed.[72] Thus, if it is desired to reverse or unwind a transaction, the transaction will need to be undertaken again, literally, in reverse.[73]

There are also elements of a DLT system that are configurable to suit the desired design and the application to which the system is to be put.[74] The configurable features include permissioning, referring to whether a system is open for anyone to join (that is, it is public or permissionless), or is private, or at least, is set up by a collaboration of parties, so that only trusted or vetted participants can partake in the control and maintenance of the system;[75] proof of work, which is a means to achieve consensus in a permissionless system;[76] 'smart contracts', referring to transactional terms and conditions embedded in computer code, which allow automatic execution of the relevant transaction once precise conformity with those terms and conditions has been estab-

---

[71] Adrian Jackson, Ashley Lloyd, Justin Macinante, and Markus Hüwener, 'Networked Carbon Markets: Permissionless Innovation with Distributed Ledgers?' (July 4, 2017), 7 <https://ssrn.com/abstract=2997099> accessed 09/10/17.

[72] Ibid, Table 1. As to immutability, see next section.

[73] There will, of course, be implications of this if, for example, the counterparties' positions have changed in the interim.

[74] Fn.71 (Jackson et al.) 8.

[75] Ibid, Table 2; also fn.35 (ASTRI); for a comparison of relative strengths and weaknesses of permissioned, unpermissioned and hybrid blockchains, see: fn.36 (Ferrarini et al./ADB 2017) 2–6; for advantages of private over public blockchains, see: Vitalik Buterin, Public and Private Blockchains, Coindesk website, 7 August 2015 <https://www.coindesk.com/vitalik-buterin-on-public-and-private-blockchains/> accessed 02/02/18.

[76] As the proposal set out in this chapter is for a permissioned system, proof of work is not considered in any detail, but rather other consensus mechanisms will be considered.

lished;[77] and arrangements for settlement, exchanges, or payment systems, which may be required in some shape or form to provide for the actual transfer of money, or settlement of physical assets, between counterparties.[78]

Configuration of all of these elements can add up to very different outcomes. For instance, the contrasting nature of the Ethereum platform, compared with the Corda™ platform: the former is public, anonymous, token-based, relies on proof-of-work for consensus, holds data on all nodes and is blockchain-based; whereas the latter is private, identity-based, tokenless, relies on proof-of-authority for consensus (notaries are used as trusted entities), data is private to the parties to a transaction and it is not blockchain-based, but blockchain inspired: yet both these platforms facilitate peer-to-peer smart contract transactions.[79]

(v)   Use case of the proposal

In these circumstances, the importance of specifying the configuration of (or, at least, the options for) the use case proposed by this book is evident, for two reasons. First, the way in which the use case is configured will determine whether the perceived benefits of the technology (considered in the next section) are actually realizable, or exist only in theory. Second, the design of the technology platform will indicate how the application should be regulated and the institutional framework required, as considered in later chapters.

The specific application of DLT proposed connects the carbon markets (that is, the emissions trading schemes (ETSs)) of individual jurisdictions that choose to participate in the network, in order to provide for inter-jurisdictional trading of their carbon assets (the units traded in the respective ETSs). Hence, the aim is to facilitate, as with the two platforms mentioned above, smart contract-based transactions peer-to-peer, in this case, across jurisdictions. For

---

[77]   Fn.71 (Jackson et al.) 8, Table 2; also Justin D. Macinante, 'A Conceptual Model for Networking of Carbon Markets on Distributed Ledger Technology Architecture' [2017] *CCLR* 243, 251. The original formulation is: 'A smart contract is a computerized transaction protocol that executes the terms of a contract. The general objectives of smart contract design are to satisfy common contractual conditions (such as payment terms, liens, confidentiality, and even enforcement), minimize exceptions both malicious and accidental, and minimize the need for trusted intermediaries. Related economic goals include lowering fraud loss, arbitration and enforcement costs, and other transaction costs'. Nick Szabo, 'Smart Contracts' (1994) <http://www.fon.hum.uva.nl/rob/Courses/InformationInSpeech/CDROM/Literature/LOTwinterschool2006/szabo.best.vwh.net/smart.contracts.html> accessed 26/01/18.

[78]   Ibid (Jackson et al.) Table 2.

[79]   Fn.56 (Simpson) slide 39. The notary design utilises a trusted authority and consensus is reached on an individual transaction basis, rather than in blocks of transactions, with limited information sharing, see fn.31 (Bech, Garratt/BIS) 58, Box A.

the market system proposed, a primary element is that it will be comprised of multiple nodes (whether each and every node would need to hold a copy of the full ledger, will be a matter of design). There would be encryption, for instance, using public/private keys and there would need to be a consensus mechanism for updating the ledger. If this updating is accumulative, such that new entries to the ledger followed consecutively on earlier entries (whether in blocks, or otherwise, being another design question), without changing or altering them, then the four key elements that identify a DLT system (outlined above) would be present.

As the network would connect the administrators of the respective ETSs, as well as the legal entities participating in each domestic ETS, the participants would all be identified. Thus, the ability for a legal person (presumably corporate, depending on the criteria applied for participation in their domestic ETS) to participate in cross-jurisdictional trading will depend on their authorization to participate in their domestic ETS. Accordingly, the distributed ledger would not be anonymous, nor public/permissionless, in the sense that anyone at all can participate. Strictly speaking, it would not be private either, in the sense of being closed to all but an exclusive group, since presumably any legal entity satisfying the relevant criteria could be authorized to trade in a domestic ETS. The network may best be described as public but permissioned, since the pre-condition for participation on the DL network would be that the legal entity was first authorized to trade in a participating domestic ETS.[80]

In the context of participation by Paris Agreement parties, it is assumed that mutual authorization of each other for the purposes of Article 6 would apply. There is the further consideration of whether participation by a jurisdiction in the DL network would imply all participants in that jurisdiction's domestic ETS were automatically considered to be authorized, by that jurisdiction's government, to trade inter-jurisdictionally, or if specific authorization for each individual legal entity to so trade would still be necessary to satisfy the requirement that use of internationally transferred mitigation outcomes to achieve nationally determined contributions is voluntary and authorized by participating parties.[81] This will be a matter for each individual jurisdiction to determine. For these purposes, it is assumed that if a jurisdiction agrees to join the network, then automatically, all the entities in its domestic ETS are considered so authorized.

---

[80] Also could be described as a hybrid: see Fn.36 (Ferrarini et al/ADB 2017) 4–5.

[81] UNFCCC: Draft Text on Matters relating to Article 6 of the Paris Agreement: Guidance on cooperative approaches referred to in Article 6, paragraph 2, of the Paris Agreement, Version 3 of 15 December 00:50 hrs, Proposal by the President, Annex paragraph 4(c) <https://unfccc.int/resource/cop25/CMA2_11a_DT_Art.6.2.pdf> accessed 15/01/20.

It follows that, as a public but permissioned DL, there would need to be configured a system providing for the type of permissioning granted to nodes, that is, identifying those permitted to view, and those permitted to interact with, the ledger. Legal entities, for example, might have permission to interact with the ledger by submitting transactions for addition to it, as well as being permitted to view that part of the ledger pertaining to their own holdings and transactions. Further, as suggested above, they might not hold a copy of the entire ledger, as this could lead to scalability problems as the ledger grows in size,[82] but might only need hold that part relating to their own holdings and transactions.[83] ETS administrators might be restrained from interacting with the ledger in the sense of submitting transactions, but might have broader viewing permission rights, for instance, by being able to view the accounts of all legal entities in their own ETS and some components of the information held on the overall ledger more generally (although perhaps not, for instance, information pertaining to individual legal entities from other jurisdictions). Consideration would also need to be given to the extent of public access to information on the ledger.

Related to this would be the consensus mechanism by which new transactions are entered on the ledger. This might operate on a distributed basis[84] but only between the administrator nodes. For example, the administrator of the ETS from which a transaction originates would perform the role of validator by confirming that the seller in the transaction is the true owner of the carbon assets being sold. They would then broadcast the information concerning that transaction (and any other transactions originating from its ETS at the same time), as other administrators would also do concerning transactions originating in their respective ETSs at that time. These validating nodes would then agree (by a mechanism they would have determined in advance) which of the transactions – presumably all, if they had all been confirmed as being correct – would be included in the block to be added to the blockchain, if the platform were to be blockchain-based, or otherwise stored one after the other in a continuous manner (but not in blocks).

---

[82]   There is a discussion of this issue in the Ethereum white paper: Ethereum, 'A Next-Generation Smart Contract and Decentralized Application Platform', White Paper <https://github.com/ethereum/wiki/wiki/White-Paper> accessed 13/02/18; scalability limitations have been identified as a weakness of permissionless DLs: see fn.36 (Ferrarini et al./ADB 2017) 3.

[83]   Richard Gendal Brown et al., 'Corda: An Introduction', White Paper, August 2016 <https://docs.corda.net/_static/corda-introductory-whitepaper.pdf> accessed 12/02/18; Richard G. Brown, 'Introducing R3 Corda™: A Distributed Ledger Designed for Financial Services', blog post, 5 April 2016.

[84]   Fn.35 (ASTRI) 10–15 provides a description of this process.

As this proposal concerns the conduct of transactions between jurisdictions, it presumes there will be contracts setting out the terms and conditions on which those transactions have been agreed. Such terms and conditions could be standardized for all transactions across the network, with provision for variable factors – parties, quantity, price, origin, mitigation value, or any other variable characteristics – to be inserted. This will be the function of smart contracts, which would allow automatic execution of the relevant transaction to which they pertain once precise conformity with the terms and conditions had been established. In conjunction with execution of the smart contract for a transaction, in order to complete the transaction, arrangements for financial settlement coordinated with the transfer of the carbon asset, will need to be in place. This aspect is considered in more detail in the next chapter.

Finally, it is envisaged that the DLT application being proposed here may, or may not, operate as a blockchain. Hence, the technology platform will continue to be referred to by the broader descriptive term, DLT, or DL, unless the context requires specific reference to a blockchain mechanism.

SPECIFIC CHARACTERISTICS OF THE PROPOSED TECHNOLOGY PLATFORM

The last sub-section introduced DLT and the use case proposed here. This section now explores in more detail the appropriateness of DLT for the proposed market. To analyze the claimed beneficial characteristics of a DL platform in the market proposed, an obvious approach would be to compare the proposal with a similar existing market based on current technology, that is, on a traditional, centralized database platform. However, as no trading network or market, such as that which is proposed, exists at present, this is not feasible. The ITL is probably the closest comparable example, but as it relates to the homogeneous trading under the Kyoto Protocol and does not provide a trading platform, is not considered appropriate for this purpose.

Comparisons might be drawn, alternatively, between the networked market proposed and the IT platforms of existing linked arrangements between jurisdictions (EUETS-Switzerland, California-Quebec). However, on one hand, it is possible the technological aspects of those comparisons would be obscured by the parallel networking-linking distinctions between the two approaches, while on the other, if it were possible for the respective approaches to be stripped back to just the IT platforms, such an exercise in comparative analysis of the relative technical IT specifications would be too removed from the legal and policy analysis the purpose of this book. Hence, this approach also is not pursued.[85]

Rather, another alternative is applied. As stated in the preceding section, the proposal can be viewed as proceeding down two independent, but interrelated arms, the first aiming to facilitate and stimulate an inter-jurisdictional market, so that it operates efficiently, encourages private sector engagement, promotes a stable carbon price and fosters the effective application of carbon finance, while the second promotes the objectives of climate policy, evidenced by the decisions of the parties at COP 21 and the terms of the Paris Agreement. Accordingly, the approach taken to analyzing the claimed benefits of DLT in the case of NCM is similarly twofold: first, by considering the DLT application in terms of the beneficial characteristics it purports to bring to financial markets (this application being probably the most extensively examined area of application for DLT at the present time); and second, by analysing the DLT application to networking carbon markets in terms of the extent to which it can better facilitate matching the requirements and expectations of the Paris Agreement, than otherwise might be achievable (that is, without DLT).

1.   DLT Application to NCM as a Financial Market

The multifarious applications of DLT,[86] particularly in relation to the financial sector, are increasing all the time, as are claims extolling the superiority of the technology over legacy systems for existing applications, or the beneficial features of new applications made possible by the technology.[87] For instance, the Chief Scientific Advisor to the UK government has stated:

> Existing methods of data management, especially of personal data, typically involve large legacy IT systems located within a single institution. To these are added an array of networking and messaging systems to communicate with the outside world, which adds to the cost and complexity. Highly centralised systems present a high cost single point of failure. They may be vulnerable to cyber-attack and the data is often out of sync, out of date or simply inaccurate.[88]

---

[85]   All the same, references and comparisons are made to the IT platforms used under current arrangements, where appropriate.

[86]   See section preceding; see also fn.34 (Siebold and Samman/KPMG) Figure 4 DLT Landscape.

[87]   Fn.28 (World Economic Forum); DTCC Connection, 'Eight Key Features of Blockchain and Distributed Ledgers Explained', 17 February 2016. <http://www.dtcc.com/news/2016/february/17/eight-key-features-of-blockchain-and-distributed-ledgers-explained> accessed 15/02/18, which sets out eight key capabilities that it claims has created the innovative platform that has the potential to modernise the post-trade financial ecosystem.

[88]   Fn.24 (Walport) 6.

DLs, on the other hand, are inherently harder to attack, the technology is resistant to unauthorized change or malicious tampering and the methods by which information is secured and updated mean that participants can share data and be confident that all copies of the ledger at any one time match each other.[89]

How the technology and its applications are perceived, however, is a question of perspective and, in this sense, the regulators can balance the picture. Thus, key challenges and possible shortcomings, of a technological nature, have been flagged to include scalability issues, interoperability with existing systems and between systems, the need for a way to settle transactions in central bank (fiat) money, the absence of a recourse mechanism for dealing with mistakes, the inability to net off positions in financial markets, and absence of scope for margin finance and short selling.[90] In relation to the governance framework, who might be permissioned in such a system, and rule design, arise as issues; privacy issues arise in relation to which parties might access what information, and regulatory and legal issues arise concerning ownership of records, liability of participants, and enforcement of obligations.[91] Key risks raised include cyber risk, fraud, and money laundering, the difficulty of identifying anomalies in such an automated system, and dealing with erroneous coding.[92]

What are the 'transformative characteristics'[93] and how do they address the issues and risks raised? According to one source, they are immutability, transparency, and autonomy.[94] Another source lists consensus, validity, uniqueness, immutability, and authentication.[95] Others emphasize the distributed ledger and consensus;[96] or the security engendered through encryption;[97] or the decentralized nature of the system across participating nodes and its peer-to-peer facility.[98]

Again, a preliminary issue is terminology, as characteristics can be described differently, depending on the perspective of the proponent. For example, anonymity, privacy, confidentiality, party identity abstraction, permissionless, trustless, security, and transparency are descriptions that might all refer to the same feature, but from different perspectives; autonomy and uniqueness might

---

[89] Ibid.
[90] Fn.35 (ESMA 2016).
[91] Ibid.
[92] Ibid. In relation to risks, see also: fn.28 (World Economic Forum).
[93] Fn.28 (World Economic Forum) uses this description.
[94] Ibid.
[95] Fn.83 (Brown et al./Corda).
[96] Fn.36 (Ferrarini et al./ADB)
[97] Freshfields Bruckhaus Deringer LLP, What's in a blockchain? <https://www.freshfields.com/en-gb/our-thinking/campaigns/digital> accessed 06/02/18.
[98] Fn.56 (Simpson).

refer to the same thing, which others might refer to as party identity abstraction. In these circumstances, for the purpose of this analysis, a set of features is selected, then by examining each in turn, consideration is given to the extent to which they afford the benefits claimed, any risks to which they give rise and how they mesh with other elements. The selected features comprise the following:

- immutability (includes traceability, auditability, robust accounting);
- decentralized (includes smart contracts);
- distributed (includes transparency and privacy, permissioning);
- security (includes hash cryptography, consensus mechanism).

(i)   Immutability

Immutability is probably the most important characteristic that DLT brings and virtually every description of blockchain or DLT refers to it. For instance: 'Immutability is a characteristic of blockchain technology … Certain features of the blockchain concept might be relaxed … but not immutability, which remains crucial …'[99] Acceptance of immutability is implicit even in the way technological challenges are identified: for instance, the European Securities and Markets Authority (ESMA) poses the absence of a recourse mechanism as an issue for dealing with mistakes once the immutable DL records a transaction.[100]

The fact that entries to a DL are cumulative, that once added they cannot be amended or edited or tampered with, is extremely important for other claimed beneficial characteristics of DLs. If entries or transactions (that is, the information related thereto), are always added cumulatively and cannot be altered once added, then the provenance of an item transacted, such as an emission unit, is easily traceable; its current ownership and history of ownership are readily ascertained; any transactions affecting it or its validity will be apparent; and any co-benefits associated with it can similarly be identified and tracked. Consequently, accounting, auditing, and reporting are facilitated. It follows logically also that, once a legal entity has transacted and sold that emission unit, the entity will be incapable of selling the unit again, unless they have first bought it back and nothing has transpired in the meantime to affect its validity, such as surrender or cancellation. In other words, this immutability characteristic facilitates robust accounting. These are all important features for the proper and efficient operation of a financial market.

---

[99]   Fn.46 (Pilkington); also see: fn.29 (Wright and De Filippi); fn.87 (DTCC Connection).
[100]   Fn.35 (ESMA 2016) 14–15, paragraph 33.

However, as Walch points out,[101] immutability isn't all that it seems, for two conceptual reasons. First, so-called 'immutable' blockchains can and have been changed post facto; and second, even though immutability is generally used to describe all types of blockchain, there is no consensus yet on what generates this feature and whether it is present in all variations.[102] In relation to the first of these reasons, the people operating the system can always agree to go back and change the record. In the two instances Walch cites, both were public blockchains (Ethereum and Bitcoin) and the action apparent to and accepted by users.[103] An ability to go back and change the ledger would be even more likely for a private (or hybrid) DL where, by definition, it is operated by a select group. Whereas blockchain is often described as immutable, this is only the case to the extent that its human creators choose not to intervene.[104]

Another view, from the security perspective, is that both permissioned and permissionless systems are only trustworthy so long as the majority of the validators are behaving honestly: in a permissioned system, there is also the need to consider the integrity of the entity that identifies and grants credentials to consortium members. 'The sanctity of the consensus mechanism, and thus the immutability of a ledger, is only upheld by trust in an identifying agent and the safekeeping of identity credentials by participants ...'[105] Thus, security and consensus are tied to immutability, which in a permissioned system, essentially relies on the trustworthiness of the person or entity running the scheme.[106]

The second conceptual point about DL immutability is the lack of agreement as to how it arises.[107] Some ascribe it to the consensus mechanism (that is, in Bitcoin, proof of work); others to the cryptography (hash functions turn data into a trunk of random characters called 'hash': see fuller explanation under (iv) security, below); others still, to the chaining together of blocks of transactions (although this is tied to the hash process). Another perspective is that other entities will not accept transactions that try to build on a modified version of some data that has already been accepted by them, the reason being that

---

[101] Fn.36 (Walch) 722 n.35, and Part IV, 735–45.

[102] Ibid, 738.

[103] Ibid, 739.

[104] Fn.36 (Finck) 668.

[105] Peter Van Valkenburgh, Director of Research, Coin Center (NFP DLT research and advocacy center), Comments to the European Securities and Markets Authority on its Consultation on Distributed Ledger Technology Applied to Securities Markets, September 2016, Coin Center <https://coincenter.org/files/2016-09/coin-center-letter -to-esma.pdf> accessed 02/02/18.

[106] Ibid.

[107] Fn.36 (Walch) 741 et seq.

transactions commit to the outcome of prior transactions, blocks to previous blocks.[108]

Ultimately, in the case of the public permissioned DL envisaged for the financial market proposed here, immutability, or more appropriately, the permanence and accuracy of the ledger record as it accumulates, will be a function of the operators, who will be the ETS administrators (hence parts of the respective governments) of the participating jurisdictions. The risk of improper collusion on their part to alter the ledger, in the first instance, will be a function of who they are and how many.[109] Additionally, once transactions have been entered and form the starting point for subsequent transactions, it would be difficult to alter the ledger without the awareness (and concurrence, one imagines) of all affected participants (that is, counterparties to the relevant transactions). If public/private key cryptography applies and, assuming the DL is a blockchain, the blocks are chained with hash functions related to the information content in the blocks, then even more so it would be necessary to engage all participants in order to alter the record.[110] This is very unlikely, given the nature of the envisaged application.

(ii)   Decentralized

One of the features that defines a DL system is a decentralized infrastructure, meaning the system is composed of multiple entities or nodes, each holding a copy of the full ledger, obviating the role of the central ledger holder.[111] While in the scheme design envisaged by this proposal, not every node may hold a full copy of the ledger, the decentralized nature of the system means there will be multiple participants capable of interacting with each other, peer-to-peer, rather than through a central body. Some commentators describe permissioned (private or hybrid) ledgers as more centralized,[112] but it has been noted that these and other design choices need to be tailored to the specific goals pursued in the particular use case.[113]

If truly decentralized, then no one entity owns the network completely. If a group of nodes control the network, however, why not just use a centralized database (as at present)? A number of reasons have been suggested by

---

[108] Fn.83 (Brown, April 2016).
[109] Also the institutional supervision proposed in the model should make improper activity highly unlikely.
[110] Fn.36 (Walch) 738–9: see discussion of Bitcoin and Ethereum 'forks'.
[111] Fn.71 (Jackson et al.).
[112] See, for instance: fn.46 (Pilkington); however, this really relates to the consensus mechanism, and they would still be decentralized in the sense of operating peer-to-peer.
[113] Fn.36 (Ferrarini et al./ADB 2017) 2.

Simpson and Wang:[114] first, once deployed, the technology is very resilient and it is very difficult to close down the entire system; if one node loses data due to, for instance, hacking or loss of power, it can recover from another node on the network; second, in an open market a middleman with a big enough platform controls pricing and will have an incentive to increase prices, whereas in a consortium everyone should have an interest in reducing transaction costs; third, if nodes can interact peer-to-peer, they are not dependent on the central party for the speed of the interaction; and nodes might control their own dataset and determine who can see it, although this will be a function of the DL design and permissioning arrangement established.

A further element of a decentralized infrastructure is the means by which nodes transact, in other words, the smart contract arrangements put in place as part of the system design. Smart contracts are not a defining element, but rather configurable to suit the desired design and the application to which the DL system is to be put.[115] They have been described as a computer code or protocol that automates the execution of certain terms and conditions of an arrangement.[116] They replicate legal contracts by coding the underlying agreement in computer language 'and have the advantage of low contracting, enforcement, and compliance costs'.[117]

Smart contracts enable transactions between counterparties digitally without the need for a trusted central counterparty, on the basis that once all the pre-conditions on the respective parties have been satisfied, the contract executes automatically. The advantage they bring, therefore, is reduced time and cost for carrying out transactions.[118] Smart contracts differ from established forms of automated contract execution of an underlying agreement, such as automated banking payments, or standing orders, in a number of respects:[119] third parties usually retain control over the transaction with those established forms, whereas the smart contract is neither administered nor controlled by a third party and, with the former, the computer program is usually run on the third party's server, ensuring their internal control; further, with traditional automated contract execution the code is exclusively in the hands of the third party responsible for it, whereas with smart contracts the DL enables all par-

---

[114] Fn.56 (Simpson) slides 49–56.

[115] Fn.71 (Jackson et al.).

[116] Fn.36 (Finck) 670–1; fn.29 (Wright and De Filippi) 10; for original formulation: fn.77 (Szabo). See also fn.77.

[117] Ibid (Finck).

[118] Fn.28 (World Economic Forum); see also fn.34 (Allens Lawyers); fn.42 (Davis et al./Linklaters).

[119] Freshfields Bruckhaus Deringer LLP, 'What's in a smart contract?' 5 February 2018 <www.lexology.com> accessed 06/02/18.

ticipants to be running the same code on a decentralized basis, enabling the peer-to-peer transaction basis.[120]

All the same, the need for third parties may not be totally obviated, with roles continuing in relation to, for instance, technical governance matters such as maintaining the technical code, auditing it against legal code, or in managing operation of the system more generally, by providing validation, ensuring regulatory compliance, and carrying out reporting functions or dealing with mistakes, errors or fraud.[121] Other than in terms of the validation role, however, the third party would, most likely, not be central to or interposed between counterparties to a transaction. As such, these functions are more properly considered as systems maintenance and management. The validation role, on the other hand, relates to the consensus mechanism design for a private DL. Many of the potential issues raised, such as insolvency of a counterparty, or payment failure, or privacy and confidentiality, can be addressed through system design (considered in chapters following).

(iii)   Distributed

To some extent, this could be seen as covering similar ground as the decentralized elements above. However, whereas the emphasis in decentralization relates to the interaction of the nodes in the network (for example, through smart contracting) and disintermediation of central third-party gatekeepers, the distributed element emphasizes the informational side – how the ledger is held, viewed, and updated. It deals with issues of consensus and permissioning, hence transparency, privacy, and confidentiality.

In a fully distributed ledger, the complete record of transactions, such as rights to payment, or ownership of an asset, would be shared contemporaneously, more or less, across the network with all participants. Thus the record is held on all nodes with concurrent updating once the correct version of the record is established. There is no one central administrator or database. For an external cyber attack to impact a distributed system, it would therefore need to infiltrate multiple copies of the ledger, not just a single central record.[122]

To establish the correct version of the record, there will need to be a consensus mechanism.[123] In the case of unpermissioned, or open ledgers, this is most

---

[120]   Ibid.

[121]   Fn.34 (Allens Lawyers).

[122]   Fn.42 (Davis et al./Linklaters).

[123]   Fn.83 (Brown, April 2016): 'The first, and most important, feature of blockchains…'; fn.34 (Siebold and Samman/KPMG): 'Consensus mechanisms are central to the functioning of any blockchain or distributed ledger'.

often a mechanism known as 'proof of work'[124] as applied in Bitcoin, although since 2012 there has also been the 'proof-of-stake' mechanism, which requires fewer calculations and is therefore less energy intensive.[125] As DLT applications in financial markets will likely be permissioned[126] and as the application proposed is also permissioned, the unpermissioned consensus mechanisms are not considered further, other than to note that the scalability limitation, related to their excessive energy demand,[127] is not the case with permissioned consensus mechanisms.

An outline of a permissioned consensus mechanism is set out earlier in relation to the use case of the proposal.[128] The main difference between the unpermissioned and permissioned DL is the degree to which it is distributed, or decentralized. The permissioned system, being applied to a limited number of nodes, would be more susceptible to cyber attack, yet simply reverting to a centralized database would only increase that risk. The permissioned system also has the advantage of being able to tailor the permissioning rights to the requirements of the participating nodes. Thus, rather than all nodes being able to view the ledger in its entirety, but without the identity of transaction participants being known (as in an anonymous, unpermissioned system), the permissioned system might be configured to allow differing levels of access to information on the ledger and differing rights to interact with the ledger, for example, by submitting transactions for adding to it. Confidentiality and privacy aspects, therefore, could be balanced with regulatory transparency needs as part of the system design.

(iv)   Security

Security in DL systems relates, primarily, to the implementation of cryptographic techniques, of which there are multiple examples.[129] One illustration

---

[124] Fn.34 (Siebold and Samman/KPMG): this was the first such mechanism, developed in 1999, and requires the system's users to repeatedly run algorithms to a mathematically complex problem; fn.35 (ASTRI) addresses the technology as does fn.36 (Ferrarini et al./ADB 2017) who also list technical references. It is not considered necessary to delve in any detail into the technical aspects by which the unpermissioned consensus mechanisms function for the purposes of this book.

[125] Fn.34 (Siebold and Samman/KPMG).

[126] Fn.35 (ESMA 2016) 8, paragraph 3.

[127] Fn.36 (Ferrarini et al./ADB 2017) 3.

[128] Preceding section; see also fn.35 (ASTRI) 10–15.

[129] For example: public key infrastructure (PKI), cryptographic hash function (HASH), secure hash algorithm (SHA-256), zero-knowledge Succinct Non-Interactive Arguments of Knowledge (zk-SNARK), homomorphic encryption (HE), Elliptical Curve Cryptography (ECC), Elliptical Curve Digital Signature Algorithm (ECDSA), software guard extensions (SGX).

is hash cryptography,[130] which applies a one-way mathematical function to summarize the relevant data as a piece of unique, fixed-size, short data called its hash value. An alteration to the data causes the hash value to change, making it impossible to decipher the original data from the hash. A change to the content of a block in a blockchain also causes the value of its hash link to change.[131]

The security offered by cryptography operates at the micro level, while security can be viewed also at a macro level, through design.[132] Design is evident in the three preceding elements: the permanent (but perhaps not totally immutable), accumulative nature of the ledger; the decentralized nature of participating nodes, transacting peer-to-peer, not needing central trusted counterparties in order to add transactions to the ledger; and the ledger held by the nodes on a distributed basis, updated by a consensus mechanism and viewed on the basis of defined permissions. The design of the overall DL, through these elements, thus contributes to security. Nevertheless, permissioned DL systems are still potentially exposed to threats including, for instance, cyber attacks, that may cause network fragmentation or performance issues. The potential for these sorts of events points to the need not only for design that avoids vulnerabilities such as network 'choke points', where the entire network can be impacted by an attack on a single node, but also to administration design, to provide for continuity in spite of such events.

2.   DLT Matching the Expectations of the Paris Agreement

While the preceding subsection considers the beneficial characteristics DLT might bring to financial markets, this subsection examines the extent to which it facilitates matching the requirements of the Paris Agreement. The signatory parties (Parties) having committed to prepare, communicate, and maintain successive nationally determined contributions (NDCs),[133] the Paris Agreement provides encouragement for carbon markets by recognizing that Parties may engage voluntarily in cooperative approaches involving the use of internationally transferred mitigation outcomes (ITMOs) towards their NDCs.[134] Voluntary cooperation in implementing NDCs is to allow for higher ambition by the Parties choosing so to act in their mitigation and adaptation

---

[130] SHA-256 is a common example.
[131] Fn.35 (ASTRI) 23.
[132] Ibid, 18.
[133] Article 4, paragraph 3, Paris Agreement.
[134] Article 6, paragraph 2, Paris Agreement. As the proposal is being outlined in terms of the connecting of ETSs, to avoid confusion, analysis of the terms of the Paris Agreement (e.g., re ITMOs) will not include the mechanism under Article 6, paragraph

actions, and to promote sustainable development and environmental integrity.[135] Additionally, where engaging in such approaches that involve the use of ITMOs towards their NDCs, the Parties shall promote sustainable development and ensure environmental integrity and transparency, including in governance, and apply robust accounting to ensure, inter alia, the avoidance of double counting.[136]

This subsection examines the proposition that the application of DLT provides innovative solutions in two areas of critical importance if operationalization of Article 6, paragraph 2 is to engage the private financial sector in building a cross-jurisdictional carbon market. They are, first, in data (information) sharing and management; and second, in transaction management.

(i)   Centrality of information sharing and management
Interpretation and implementation of Article 6, paragraph 2, is subject to the guidance being developed by the Subsidiary Body for Scientific and Technological Advice (SBSTA), in accordance with the decision of the Parties in adopting the Paris Agreement.[137] All the same, it is observed that the international transfer of mitigation outcomes for use towards Parties' NDCs sits in a matrix of inter-related principles, or themes, infused throughout the decision of the Parties and in the Paris Agreement.[138] In this matrix, the centrality of information and information management, to all aspects of the Paris Agreement relating to mitigation, is an unavoidable conclusion. Information, principally in relation to its provision by Parties is key, as underscored by concepts such as transparency, robust accounting, and reporting. These three concepts are now considered.

(a) Transparency
Transparency has been described as having become 'one of the fundamentally distinctive traits of contemporary Western culture', and that its opposites, 'such as secrecy and confidentiality, have taken on a negative connotation'.[139]

4, as this is project crediting-based, as opposed to allowance based. See also clarification of approach in Chapter 1.

[135]   Article 6, paragraph 1, Paris Agreement.

[136]   Fn.134 (Art.6. para.2).

[137]   Decision 1/CP.21, FCCC/CP/2015/10/Add.1, 29 January 2016, paragraph 36 <http://unfccc.int/resource/docs/2015/cop21/eng/10a01.pdf> accessed 13/03/17.

[138]   Note that in reviewing this matrix of themes, the focus and emphasis is on mitigation.

[139]   A. Bianchi, 'On Power and Illusion: The Concept of Transparency in International Law' in A. Bianchi and A. Peters (eds.), *Transparency in International Law* (Cambridge University Press, 2013) 1–2.

It 'is not immediately associated with international law',[140] which classically has been a device to 'formalise the outcomes of inter-State negotiations on select issues of mutual concern'.[141] However, this paradigm has proved to be incomplete in relation to international environmental law, where 'environmental concerns implicate individuals and groups in society, not just states'.[142] In particular, transparency is not a new concept in climate policy, having been recognized as a principle or good practice since adoption of the UNFCCC, an early application being in the National Communications thereunder.[143]

While it is 'often associated with information and knowledge, legitimacy and accountability, participatory democracy and good governance', transparency 'means different things to different people in different contexts'.[144] For instance, one author defines it as '… a system in which the relevant information is available … now widely seen as an important element of institutional legitimacy, both for global institutions and national authorities'.[145] Yet this description begs further questions, such as relevant to whom? Relevant when? Relevant for what purposes? And does the institutional legitimacy mentioned derive from procedural transparency (e.g., openness in the process)[146] or from transparency in the outcomes?[147]

PARIS AGREEMENT AND TRANSPARENCY   In Paris, the Conference of the Parties (COP) adopted decisions in relation to transparency of action and support,[148] including establishing a Capacity-building Initiative for Transparency,[149] and agreeing to establish an enhanced transparency framework for action and support, taking account of Parties' different capacities.[150] This framework is to be implemented in a facilitative, non-intrusive, non-punitive manner, respect-

---

[140] Ibid, 3.

[141] J. Brunnée and E. Hey, 'Transparency and International Environmental Institutions' in A. Bianchi and A. Peters (eds.), *Transparency in International Law* (Cambridge University Press, 2013) 26.

[142] Ibid.

[143] Sina Wartmann and Raúl Salas, Ricardo Energy & Environment; Daniel Blank, GIZ, 'Deciphering MRV, accounting and transparency for the post-Paris era', January 2018, GIZ, <https://www.transparency-partnership.net/system/files/document/MRV .pdf> accessed 12/02/18.

[144] Fn.139 (Bianchi) 8.

[145] Anne-Sophie Tabau, 'Evaluation of the Paris Climate Agreement according to a Global Standard of Transparency' [2016] *CCLR* 23.

[146] Fn.141 (Brunnée and Hey) 25: what these authors refer to as 'transparency of governance'.

[147] Ibid: what these authors refer to as 'transparency for governance'.

[148] Fn.137 (Decision 1/CP.21) paragraphs 84–98.

[149] Ibid, paragraph 84. See also <https://www.cbitplatform.org/>.

[150] Article 13, paragraph 1, Paris Agreement.

ful of national sovereignty and avoiding placing undue burden on Parties.[151] The purpose is to provide a clear understanding of climate change action, in light of the objective in Article 2, UNFCCC, including clarity and tracking progress towards achieving NDCs.[152] According to Tabau, the aim:

> ... is not, as was the case with the Kyoto Protocol, to link transparency of implementation and compliance, but rather to enhance trust in order to raise ambition ... by generating forward-looking and real time information, this transparency framework will dissipate fears of free-riding and competitive disadvantage, allow mutual learning and support, and send a signal beyond the level of States, in particular to private investors.[153]

Thus, as perceived by Tabau, an outcome of the trust generated by greater transparency, would be engagement with the private sector.

Each Party is to regularly provide a national inventory report of anthropogenic emissions by sources and removals by sinks,[154] and the information necessary to track progress in achieving its NDC.[155] In accounting for anthropogenic emissions by sources and removals by sinks corresponding to their NDCs, Parties agree, inter alia, to promote transparency,[156] to do so as well in communicating their NDCs,[157] and to ensure transparency in using ITMOs towards their NDCs.[158] The COP, serving as the meeting of the Parties to the Paris Agreement (CMA) will adopt common modalities, procedures and guidelines for the transparency of action and support at its first session[159] and, to this end, the Ad Hoc Working Group on the Paris Agreement (APA) has been requested to develop such, inter alia, taking into account the importance of facilitating improved transparency over time,[160] and the need to promote transparency, accuracy, completeness, consistency, and comparability.[161]

'The word "transparency" appears 30 times in the text of the decision adopted by the COP 21 and 13 of these occurrences are contained within

---

[151] Article 13, paragraph 3, Paris Agreement.

[152] Article 13, paragraph 5, Paris Agreement.

[153] Fn.145 (Tabau) 30.

[154] Article 13, paragraph 7(a), Paris Agreement.

[155] Article 13, paragraph 7(b), Paris Agreement.

[156] Article 4, paragraph 13, Paris Agreement. Also, in this context, they commit to promote accuracy, completeness, comparability, and consistency of the information.

[157] Article 4, paragraph 8, Paris Agreement and Decision 1/21, paragraph 13.

[158] Fn.134 (Art.6, para.2).

[159] Article 13, paragraph 13, Paris Agreement. The first session of the CMA (CMA.1) was at COP 24, December 2018.

[160] Fn.137 (Decision 1/CP.21) paragraph 92(a).

[161] Ibid, paragraph 92(c).

the Paris Agreement'.[162] Given the pervasive references, the Initiative, and the framework, it is understandable that transparency has been described by some commentators as being 'core to the whole treaty'.[163] And while these processes should be positive for information sharing and management, in themselves they provoke further questions about transparency. For instance, notwithstanding references in the decisions taken in Paris[164] it is not clear the scope for non-state actors' (civil society) participation or contribution to the development of the Initiative and framework, or in the work of the APA. Thus, it is not clear whether there is transparency of process such as to afford, for instance, the APA with the 'institutional legitimacy' referred to above.[165]

Two further observations are made concerning transparency under the Paris Agreement. Modalities, procedures, and guidelines (MPGs) for the transparency framework were adopted at CMA.1.[166] The purpose of the framework is expressed as being '… to provide a clear understanding of climate change action in the light of the objective of the Convention as set out in its Article 2 …'[167] while the guiding principles of the MPGs include 'Promoting transparency, accuracy, completeness, consistency and comparability'.[168] Notwithstanding the stated purpose as being to provide understanding of climate action, although not explicitly, the transparency provisions suggest an emphasis on the provision, rather than both provision and receipt, of information. Yet receiving the information (Who? How? When?), interpreting the information (importing the notion of capacity to do so) and the ability to understand what it means

---

[162] Fn.145 (Tabau).

[163] Lambert Schneider et al., 'Environmental Integrity under Article 6 of the Paris Agreement' Discussion Paper, March 2017, German Emissions Trading Authority (DEHSt) at the German Environment Agency, 24. <https://www.dehst.de/SharedDocs/ Downloads/EN/JI-CDM/Discussion-Paper_Environmental_integrity.pdf?__blob= publicationFile> accessed 28/03/17.

[164] Fn.137 (Decision 1/CP.21) paragraphs 133–6.

[165] Fn.141 (Brunnée and Hey). For instance, in its work developing guidance on cooperative approaches referred to in Article 6, paragraph 2, of the Paris Agreement, the SBSTA has open sessions, which clearly imports that it must also have closed sessions, raising questions about the transparency of its processes.

[166] Decision 18/CMA.1, Report of the Conference of the Parties serving as the meeting of Parties to the Paris Agreement on the third part of its first session, held at Katowice from 2 to 15 December 2018, Addendum, Part two, FCCC/PA/CMA/2018/3/ Add.2, 19 March 2019, Action taken by the Conference of the Parties serving as the meeting of Parties to the Paris Agreement, Decisions adopted by the Conference of the Parties serving as the meeting of Parties to the Paris Agreement, <https://unfccc .int/sites/default/files/resource/cma2018_3_add2%20final_advance.pdf>   accessed 07/05/19.

[167] Ibid, paragraph 1.

[168] Ibid, paragraph 3(d).

for the recipient, what it means in the particular context and how to use that knowledge, surely need to be equally balanced with the provision of it, given the importance of the concept of transparency and for building trust?

Second, Article 13 provides, inter alia, that the enhanced transparency framework for action and support is to be implemented in a facilitative, non-intrusive, non-punitive manner, respectful of national sovereignty. The MPGs repeat this formulation as part of the first guiding principle '... implementing the transparency framework in a facilitative, non-intrusive, non-punitive manner, respecting national sovereignty ...'[169] and again in setting out how technical expert review should be implemented.[170] What it means in that last context might be gleaned from the following paragraph, which states that expert technical review teams, inter alia, shall not review the adequacy or appropriateness of NDCs, or of a Party's domestic actions.[171] While it is the prerogative of the CMA to put boundaries around the processes it establishes, the inference might be that investigative analysis should be limited in scope. Relating this back to the preceding point, the further question is whether this is intended to or will, in fact, also have a dampening effect on the ability to interpret information that flows from this enhanced transparency framework. This is pertinent to the networked market proposed, to which consideration is now given.

PROPOSED MARKET MODEL AND TRANSPARENCY   A number of the elements of the model proposed are directed to ensuring both transparency and the trust it is hoped to generate. First, the ledger can be distributed to all nodes, so that, theoretically at least, any participant can see all transactions. In reality, which parties can view the ledger and which can interact with it will be a function of the levels of permissioning accorded them. Nevertheless, the very fact that information could be there, on all or at least a specific number of the nodes, would increase, it is suggested, the onus on jurisdictions participating.

Second, more detailed information on the units transferred can be held with the distributed ledger, in contrast to the limited accounting role performed by registries to date.[172] The Kyoto Protocol's ITL has been described as opaque, despite being public, only recording transactions between countries, not indi-

---

[169]   Ibid, paragraph 3(a).
[170]   Ibid, paragraph 148.
[171]   Ibid, paragraph 150(a)–(e).
[172]   National registries hold no price or contract information, perform no role in relation to trading or clearing and hold no information concerning emissions: Anthony Hobley and Peter Hawkes, 'GHG Emissions and Trading Registries' in David Freestone and Charlotte Streck (eds.), *Legal Aspects of Implementing the Kyoto Protocol Mechanisms: Making Kyoto Work* (Oxford University Press, 2005) 128–9.

vidual account holders and not holding any information about contracts for forward delivery or options.[173] It is noted also that the European Commission has been criticized for a lack of transparency in aspects of its implementation of the EUETS.[174]

Third, the accuracy of information concerning transactions will be ensured by elements such as the consensus mechanism between participating ETS administrators, the accumulative nature of the ledger and oversight of the market operation by supervisory bodies (elaborated in later chapters). Again in contrast, it might be noted that concerns were raised by the European Court of Auditors in its 2015 report on the EUETS pertaining to suitability of the supervisory framework for the emission allowance market. They included the need for effective cooperation between European Commission entities responsible, on the one hand, for the carbon market and, on the other, for financial market regulation. No EU level oversight of the emissions market existed, no price or financial information relating to transactions was recorded, and there were no integrated procedures for investigating suspicious transactions.[175] Systems related to the EU Registry for processing fundamental information also showed certain shortcomings, for example, procedures for opening accounts were not sufficiently robust and transactions were insufficiently supervised and monitored at the EU level.[176]

Fourth, the permanent, accumulative nature of the distributed ledger is an element that ensures the integrity of the record. This is reflected also in the traceability of data through the ledger, thus ensuring accuracy of the record. Security can be provided, also, by the distributed nature of the ledger and by the use of encryption. Under the current UNFCCC arrangements, communications between registries and the ITL rely on functional specifications that specify use of connections using encrypted messaging over the internet, just as under the proposal. Current communications are encrypted using Secure Socket Layer (SSL),[177] which relies on certificate authorities to issue digital

---

[173] Elizabeth Lokey Aldrich and Cassandra L. Koerner, 'Unveiling Assigned Amount Unit (AAU) Trades: Current Market Impacts and Prospects for the Future' (2012) 3 *Atmosphere* 229–45, 233.

[174] Fn.40 (ECA) paragraphs 64–79.

[175] Ibid, paragraphs 14–24.

[176] Ibid. paragraphs 29–41. It is noted that in the proposed market these procedures will remain primarily the responsibility of participating jurisdictions, however, there will be additional checks and balances in the market operation rules: see next chapter.

[177] UNFCCC Secretariat, Data Exchange Standards for Registry Systems under the Kyoto Protocol, Technical Specifications (Version 1.1.11), 24 November 2013, 6–7 <https://unfccc.int/files/kyoto_protocol/registry_systems/itl/application/pdf/data_exchange_standards_for_registry_systems_under_the_kyoto_protocol.pdf> accessed 14/11/18.

certificates for identity and authentication purposes, but these have been described as not providing any actual security, as it is claimed that the average user doesn't bother to verify the certificate exchanged.[178] However, other than this observation, no comment is made on the relative technical merits of different types of cryptography, as it is not necessary for these purposes.

Finally, the independent assessment process by which the mitigation values of mitigation outcomes are assessed should afford the proposed market with another layer of transparency. Whereas the preceding elements all pertain to the functional operation of the distributed ledger itself, this element deals specifically with the interpretation placed on the information about the schemes generating the units to be traded in the market and the jurisdictions from where those schemes derive. As such, this aspect is tied directly to the information provided by Parties under the Paris Agreement and so may be influenced by the dampening effect of the 'non-intrusive, non-punitive manner, respectful of national sovereignty' qualification on implementation of the enhanced transparency framework for action and support, pursuant to Article 13. While intergovernmental processes and bodies like the technical expert review panels may be constrained, the market is more likely to respond if sentiment is that one particular jurisdiction or another is not pulling its weight. This flags a potential point of friction between the intergovernmental processes and private sector engagement through the market. However, if transparency is to build trust and, as Tabau postulates,[179] send a signal beyond the level of States to private investors, market sentiment will provide a useful independent yardstick on effort.

(b) Robust accounting

In accounting for anthropogenic emissions by sources and removals by sinks corresponding to their NDCs, Parties agree also to ensure the avoidance of double counting,[180] implying the need for robustness in their accounting processes. In using ITMOs towards their NDCs, Parties must apply robust accounting to ensure, inter alia, the avoidance of double counting.[181] The need to ensure that double counting is avoided is a consideration also for the APA.[182]

The model proposed relates principally to the interaction between participating entities. Thus, elements such as those ensuring the accuracy of information concerning transactions, security afforded by encryption and the permanent, accumulative nature of the ledger – with information added by consensus –

---

178 Fn.105 (Van Valkenburgh).
179 Fn.145 (Tabau).
180 Article 4, paragraph 13, Paris Agreement.
181 Fn.134 (Art.6, para.2).
182 Fn.137 (Decision 1/CP.21) paragraph 92(f).

should ensure traceability of provenance, facilitating an environment of robust accounting in transactions that should avoid double counting. Accounting in relation to emissions and removals corresponding to their NDCs by individual Parties participating in a network based on a distributed ledger (DL), of which their domestic registry forms a part, will nevertheless remain a function of the robustness in their domestic accounting processes. Part of the process for a jurisdiction to join the network would need to include establishing the accuracy of the domestic accounts of entities they authorize to trade as at the time of joining.

It is conceivable also that a jurisdiction may wish to maintain its existing domestic registry IT system and apply software that provides an interface to the DL network, so that legal entities regulated by it and operating on the existing legacy IT system can still trade on the network. While it is understood that this is technically feasible, such an arrangement would put that jurisdiction's domestic registry outside the DL network. Thus, elements of the network that ensure accuracy of the information, security, and robustness of the accounting processes would stop at that interface.

(c) Reporting

Reporting is a key part of measurement, reporting, and verification (MRV), which has evolved from individual requirements under the UNFCCC to become a robust framework,[183] that now forms the basis for the modalities, procedures, and guidelines for the enhanced transparency framework.[184] The centrality of information and information management is apparent not only from the significance attaching to transparency – that is, the openness, availability, and clarity of that information – and in the way it is compiled (through robust accounting), but also in terms of how it is communicated. All Parties agree, in communicating their NDCs, to provide 'the information necessary for clarity, transparency and understanding' as they have also been exhorted to do in relation to their intended NDCs.[185] The NDCs communicated by Parties are to be recorded in a public registry,[186] for which the Subsidiary Body for Implementation (SBI) is to develop modalities and procedures.[187]

---

[183] Fn.143 (Wartmann et al./GIZ) 15.

[184] Fn.137 (Decision 1/CP.21) paragraph 98; also fn.143 (Wartmann et al./GIZ) 22.

[185] Fn.157 (Art.4, para.8 and Decision 1/21, para.13). See also fn.137 (Decision 1/21) paragraphs 25, 27.

[186] Article 4, paragraph 12, Paris Agreement.

[187] Fn.137 (Decision 1/CP.21) paragraph 29. Again, transparency questions arise in relation to how the SBI will go about this process and what it will provide about the information in the public registry and access to it.

Parties also have regular inventory reporting obligations[188] in relation to tracking NDC progress,[189] and the MPGs take into account the importance of improved reporting over time.[190] Counterparties to transactions involving ITMOs are urged to report transparently, including on outcomes used to meet international pledges, with a view to promoting environmental integrity and avoiding double counting.[191] The CMA must also periodically take stock of implementation to assess collective progress towards achieving the purpose of the Paris Agreement (the 'global stocktake').[192]

So far as this proposal is concerned, ETS administrators of the participating jurisdictions will have permissioned access to the ledger, as will the entity that operates the ledger. Reporting, as it relates to transactions between jurisdictions, should be a matter of interrogating the ledger, functionality for which would be built into the design. Thus, it is anticipated that these reporting obligations would be capable of being addressed.

(ii)   Transaction management

The second area of critical importance to a cross-jurisdictional carbon market is in relation to transaction management. As noted above, provision for the international transfer of mitigation outcomes for use towards Parties' NDCs sits in a matrix of inter-related principles, infused throughout the decisions of the Parties and the terms of the Paris Agreement. These principles address the potential impact of transactions involving international transfers of mitigation outcomes expressly in terms of the need to ensure environmental integrity, and in the application of robust accounting to ensure, inter alia, the avoidance of double counting. This is especially so in provisions dealing with how Parties account for their NDCs[193] and how Parties use ITMOs towards NDCs.[194]

(a) Environmental integrity

Environmental integrity is referred to in the context of Parties engaging in voluntary cooperation in the implementation of their NDCs,[195] and to the extent that, in so doing, they use ITMOs.[196] Yet, as is the case with a number of the concepts introduced, the decisions of the Parties and Paris Agreement

---

[188]   Fn.154 (Art.13, para.7(a)).
[189]   Fn.155 (Art.13, para.7(b)).
[190]   Fn.166 (Decision 18/CMA.1) paragraphs 7–9.
[191]   Ibid, paragraph 107.
[192]   Article 14, paragraph 1, Paris Agreement.
[193]   Fn.156 (Art.4, para.13).
[194]   Fn.134 (Art.6, para.2).
[195]   Fn.135 (Art.6, para.1).
[196]   Fn.134 (Art.6, para.2).

provisions do not define what 'environmental integrity' means in this context.[197] Commentators have noted that Parties to the negotiations seem to view the concept as being confined to risks undermining GHG mitigation action, as opposed to broader environmental damage.[198] Some Parties' submissions have referred to additionality, or the avoidance of transfers of 'hot air', but none actually defines environmental integrity.[199] In the context of Article 6, some commentators understand environmental integrity to mean 'the use of international transfers does not result in higher global GHG emissions than if the mitigation targets in NDCs had been achieved only through domestic action, without international transfers'.[200] This definition may seem limiting, considering the multiple references to the expression from which a broader intended meaning might be inferred.[201] Nevertheless, it is sufficient in the context of international transfers in accordance with Article 6, paragraph 2, as considered here.

Aspects of an IT system architecture based on DLT as proposed could ensure that environmental integrity – as so defined – can be maintained when international transfers take place. For example, with DLT the basis on which a jurisdiction opts to engage in international transfers (the rules governing that jurisdiction's participation, or perhaps more accurately, governing the basis on which it authorizes an entity to engage in international transfers) can be embedded in the computer code that sets out the transactional terms and conditions on which those entities engage in transactions (the so-called 'smart contracts'). If these rules were not satisfied in a particular case, the code would

---

[197] Although an earlier draft negotiating text provided some indication in terms of the information on cooperative approaches that Parties potentially must submit: paragraphs 28(h) and (i), UNFCCC, SBSTA49: Draft Text on SBSTA 49 agenda item 11(a) Matters relating to Article 6 of the Paris Agreement: Guidance on cooperative approaches referred to in Article 6, paragraph 2, of the Paris Agreement, Version 2 of 8 December 10:00 hrs, Annex <https://unfccc.int/sites/default/files/resource/SBSTA49_11a_DT_v2.pdf> accessed 21/01/19.

[198] Fn.163 (Schneider et al./Environmental Integrity) 12.

[199] Ibid.

[200] Ibid.

[201] For instance, the Parties agree to promote environmental integrity in accounting for anthropogenic emissions by sources and removals by sinks corresponding to their NDCs (Article 4, paragraph 13); the AWGPA is requested to take into account the need to ensure environmental integrity in developing recommendations for common modalities, procedures and guidelines for the transparency of action and support (Decision 1/CP.21, paragraph 92(g)); and counterparties to transactions involving ITMOs are urged to report transparently, including on outcomes used to meet international pledges, with a view, inter alia, to promoting environmental integrity (Decision 1/CP.21, paragraph 107). As such, it is open for the expression to be interpreted to include broader environmental impacts, beyond just GHG emissions.

not permit the transaction to proceed. In this way, a jurisdiction could also set its national requirements and have scope to vary them, to adapt to changing economic, market, or environmental circumstances. There might also be more general rules embedded in the code addressing environmental integrity (for example, directed towards ensuring that a transaction could not result in higher GHG emissions by either counterparty), supplementarity and other requirements, applying to all international transfer transactions, regardless of the jurisdictions involved. Examples of these are set out in the next chapter and Appendix.

(b) Robust accounting, avoidance of double counting

Robust accounting is 'a key prerequisite for ensuring environmental integrity'[202] and 'crucial for ensuring environmental integrity and providing transparency on climate action'.[203] The SBSTA is charged with the task of developing and recommending guidance for the application of robust accounting to ensure, inter alia, the avoidance of double counting as referred to in Article 6, paragraph 2, Paris Agreement.[204] It was to complete this task by the first session of the CMA.[205]

Like 'transparency' and 'environmental integrity', what 'robust accounting' or 'avoidance of double counting' are intended to mean, is not spelt out. In order to begin defining these terms, consideration must first be given to the question of what is to be robustly accounted for, in this context. Yet the Paris Agreement provides no definition of 'mitigation outcomes', the subject of international transfers. The question of what are mitigation outcomes and related issues, such as methods for their quantification, technical tools and infrastructure for their operationalization and management, the means to

---

[202] Fn.163 (Schneider et al./Environmental Integrity) 13.

[203] Lambert Schneider et al., 'Robust Accounting of International Transfers under Article 6 of the Paris Agreement, Discussion Paper', October 2016, German Emissions Trading Authority (DEHSt) on behalf of German Environment Agency, 4 <https://www.dehst.de/SharedDocs/downloads/EN/project-mechanisms/Robust_accounting_paris_agreement_discussion_paper.html?nn=8623984> accessed 14/05/17.

[204] Fn.137 (Decision 1/CP.21) paragraph 36.

[205] Ibid. This was at COP 24, in December 2018. However, consensus could not be reached, so the decision was for the SBSTA to continue consideration and forward a draft decision for consideration at CMA.2: Decision -/CMA.1, Matters relating to Article 6 of the Paris Agreement and paragraphs 36–40 of decision 1/CP.21 <https://unfccc.int/sites/default/files/resource/auv_cp24_i4_Art.6.pdf> accessed 21/01/19. CMA.2 also failed to reach a consensus: UNFCCC CMA.2: Report of the Conference of the Parties serving as the meeting of Parties to the Paris Agreement, second session, Madrid, 2–13 December 2019, FCCC/PA/CMA/2019/L.9 <https://unfccc.int/sites/default/files/resource/cma2019_L09E.pdf> accessed 15/01/20.

ensure environmental integrity through robust accounting rules, and comparability of outcomes, are the subject of consideration by the SBSTA[206] and comment by observers.[207] Nevertheless, an IT system infrastructure based on DLT that can incorporate the transactional rules Parties agree as part of the computer code by which transactions proceed (smart contracts), to ensure that appropriate adjustments are made to the ledger automatically as part of any transactional process, should provide the robustness of accounting mandated at Paris. Adjustments to the ledger would be accumulative and more or less immutable, thereby guaranteeing traceability and permanence, facilitating auditability. The distributed nature of the ledger and constant updating by consensus of all copies, to ensure all are the same, would remove any need for reconciliations.

SUMMATION OF THE PROPOSAL

This chapter elaborates the concept of and theory for a market between carbon markets, a trading platform connecting and facilitating transactions between individual, separate markets, each of which will continue to operate as an autonomous operation in its own jurisdiction, while participating on the network created by the connection. The proposed market consists of two distinct elements, being first, the networking of the individual markets across this trading platform, and second, the platform operating on distributed ledger IT architecture. It aims to facilitate and stimulate an inter-jurisdictional market that will encourage private financial sector engagement, while at the same time promoting the objectives of climate policy evidenced in the Paris Agreement.

Networking is not current practice, presently being only conceptual in nature. The current approach for connecting carbon markets from different jurisdictions is for them to link, which involves alignment of schemes, policies, laws, processes, and so on. This gives rise to political issues, stemming from the perceived impact of system alignment on the sovereignty of the participant jurisdictions. Networking better addresses these issues, as the inherent problem of imbalance of negotiating positions would not arise. Networking also holds out a more time efficient process by avoiding the need to homog-

---

[206] See, for example, International Institute for Sustainable Development, Earth Negotiations Bulletin, Vol.12 No.747, Summary of Katowice Climate Change Conference: 2–15 December 2018, 17, 18 <http://enb.iisd.org/download/pdf/enb12747e.pdf> accessed 12/03/19.

[207] Fn.203 (Schneider et al./Robust Accounting); also see: Martin Cames et al., 'International market mechanisms after Paris', Discussion Paper, (November 2016), German Emissions Trading Authority (DEHSt) on behalf of German Environment Agency.

enize laws, systems, registries, policies, and other elements of the respective participating jurisdictions' systems. Hence, many legal and practical issues might be avoided, thereby promising a more flexible arrangement.

The global recognition that technological developments are occurring that fundamentally change how financial services are provided, how markets, business, and government operate, leads to a conclusion that in proposing a model for networking carbon markets, it is necessary and desirable to propose the technology on which the networked market platform should operate. Application of distributed ledger technology in this context is not without issues. Some identified with the technology include scalability, interoperability with existing and between systems, need for a way to settle transactions in central bank money, absence of a recourse mechanism for dealing with mistakes, and no scope for margin finance and short selling. Key risks that have been raised include cyber risk, fraud and money laundering, difficulty in identifying anomalies, and how to deal with erroneous coding.

At the same time, DLT offers useful features, including (qualified) immutability (supporting traceability, auditability, and robust accounting); decentralized participants, and so disintermediation of transaction gatekeepers (using smart contracting to facilitate transactions); distributed information sharing and management (enabling balancing of transparency with privacy, and the permissioning mechanism); and security (based on hash cryptography, and the consensus mechanism). Realization of these elements and potential benefits, resolution of issues, and management of risks, and how the application should be regulated, will be a function of use case design. In the model proposed, all participants would be identified, so the DL would be public but permissioned; and the consensus mechanism, it is proposed, should be based on nodes of the administrators from participating jurisdictional schemes (ETSs).

It has been argued in this chapter that the application of DLT provides innovative solutions in two areas of critical importance if operationalization of Article 6, paragraph 2 is to engage the private sector in building a cross-jurisdictional carbon market. They are first, in data (information) sharing and management; and second, in transaction management. As examined, elements of the decisions of the Parties, and in the Paris Agreement, concerning mitigation, so far as they relate to information and information management, emphasize transparency but appear to focus more on the provision of information.[208] Yet information is not the same as transparency. Rather transparency involves also the other side of the information coin, namely its receipt, interpretation, use, and understanding. These aspects in relation to information (that is, receipt,

---

[208] Although it is acknowledged that UNFCCC work is continuing in this respect through bodies such as APA, SBSTA, and SBI.

use, interpretation, understanding), correspond with assumptions that underlie the networked carbon markets (NCM) concept, for instance that:

- governments need information about jurisdictions and schemes with which they may consider 'connecting' their scheme;
- changes may take place in a scheme and its effectiveness over time, or with a jurisdiction's economy, so information needs to be collected and monitored on an on-going basis;
- similarly, market participants need information to make informed investment decisions;
- some governments and market participants have the resources to undertake due diligence on an ongoing basis to make these decisions, while others may not. For those that do not have access to these resources, independent sources of information are important; and
- it is implicit that governments retain sovereignty to act on this information as they see fit, and the hegemony over their schemes and policies.[209]

Information not only needs to be available, but also needs to be reliable: in terms of the Paris Agreement, there must be transparency, accuracy, completeness, consistency, and comparability of information. The various elements of an IT system architecture based on DLT would seem to be able to address these requirements. As noted, with DLT the ledger is accumulative and permanent (more or less immutable, as previously discussed), thus affording security of transactions, for instance, through fraud prevention. This can facilitate robust accounting and the avoidance of double counting, since entries are only added once they have satisfied a consensus process in which the relevant authorized entities agree they are correct and should be added to the ledger. Once in the ledger, it is very unlikely they can be altered, other than by overall consensus.

From a technical perspective, individual features and elements described as part of the DL such as the accumulative nature of the ledger, and cryptographic security, could equally well be incorporated using a centralized database. The question that needs to be considered continually is whether the distributed architecture adds anything that could not otherwise be achieved using a centralized database currently in use such as, say, the ITL, suitably adapted. The answer proffered is that consideration of the technical arrangements must be set in the broader overall context. Application of the DL, particularly for networking carbon markets, as proposed, affords greater flexibility for jurisdictions to access, or conversely leave, the networked market, according to

---

[209] Justin Macinante, 'Networking Carbon Markets: Key Elements in the Process', 2016, World Bank Group Climate Change, 9,10 <https://openknowledge.worldbank.org/handle/10986/25750> accessed 21/08/17.

their perception of domestic economic suitability, as well as a level-playing field irrespective of economic size or development.

Information needs not only to be available and reliable, it needs to be capable of interpretation and understanding. In the market proposed, this is the function of the independent evaluation process, the mitigation value assessment. The next chapter (Chapter 7) illustrates how the market proposed might be implemented in practical terms. This elaboration, in turn, provides the basis for analysis of the potential regulatory and institutional frameworks for such, in the following chapters (Chapters 8 and 9).

7. Practical implementation of the proposed networked market

Building on the preceding chapter, this chapter elaborates more tangibly how the proposed market might be implemented. The first section sets out the elements of the proposed market, before consideration is given to alternative potential market structures under Article 6 of the Paris Agreement. Elaboration of the proposal in this and the preceding chapter provides the foundation for subsequent analysis of the institutional and regulatory frameworks.

THE PROPOSED MARKET ELABORATED

The proposal is elaborated in terms of the following five areas:

- infrastructure of the market;
- rules for the operation of the distributed ledger;
- operational mechanisms that will be required, being (i) a mechanism for valuing differences between units of participant jurisdictions in terms of mitigation; and (ii) a mechanism to effect transactions;
- transactional rules as part of a regulatory framework; and
- participants, on a jurisdictional, cross-jurisdictional and supra-jurisdictional basis.

There are overlaps across these categories, as to a degree they just examine the market from different perspectives: for instance, infrastructure, categories of participant and then, the institutional framework, in the following chapter. This aims to demonstrate how the various elements will mesh together to form a coherent scheme.

1.   Infrastructure Considerations

The market can be visualized at three levels, illustrated in Figure 7.1, being in the middle, the jurisdiction where an emissions trading scheme (ETS) operates: the registry, or jurisdictional, level (the single market ledger in Figure 7.1 below); below the jurisdictional level, the entities in the ETS, trading at an intra-jurisdictional (i.e., corporate – single organization ledger) level; and

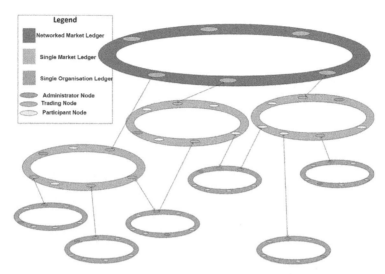

Source: Courtesy of Adrian Jackson, EPCC, University of Edinburgh.

Figure 7.1   Federated ledgers

above the jurisdictional level, the network where trading transactions (international transfers) occur: the inter-jurisdictional (networked market ledger) level. There is the possibility of sub-levels between the jurisdictional and intra-jurisdictional levels, in the case of, for instance, provincial or municipal-based ETSs operating within a jurisdiction. For the sake of simplicity, the discussion here is based on just the three: inter-jurisdictional (networked market); jurisdictional (single market – ETS/registry); and intra-jurisdictional (single organization or entity/corporate) levels.

(i)   Existing infrastructure and network design
Some of the infrastructure exists already, while some new infrastructure would be required. The starting point is that ETSs exist in various jurisdictions.[1] They consist of registries managed by administrators (who may also be the ETS regulator); entities with compliance obligations under the scheme; and other entities (such as traders and market makers) that are also authorized to trade in the ETS. Assuming that the ETS in each jurisdiction operates electronically,

---

[1]   These would operate within a domestic legal framework, being based on primary and/or secondary legislation of the jurisdiction.

the ETS administrator and each entity authorized to trade will be a node.[2] The nodes will already be capable of connecting electronically with each other, through the internet. In each ETS, the registry is the ledger and when another ETS networks with the first, their registries together become the ledger and so on, as others join the network. The ledger is in effect, therefore, a collective database for all participating registries. If the ledger electronic architecture is fully distributed, then each node will be on the distributed ledger (DL) network and operate the relevant DL software.

If each node on the network holds a full copy of the network ledger, a number of technical considerations arise. For example, this may present a problem for individual participants because of the computing and memory capacity it would require them to install, thus impacting the degree to which the network is scalable. This issue may be accounted for by design that limits the amount of historical transactions needing to be held and accessible on each and every node. Thus, authorized entities might primarily hold that part of the ledger relating to current transactions (what these include would need to be defined), while perhaps just the ETS administrator, who itself would not interact to engage in transactions, might hold the complete updated record of all ledger transactions.

Second, related to the preceding point, is consideration of how new entrants to the network get up to speed, so to speak, in terms of holding a copy of the ledger. If, as suggested in the preceding point, the ETS administrator holds the full ledger history and authorized entities trading on the network only hold that part of the ledger relating to current transactions, this consideration arises twice: first in respect of the ETS administrators and second, in relation to the authorized trading entities. The technical and timing requirements for the two situations would differ; nevertheless, it is assumed that these issues are capable of being addressed by the technical design and participating nodes' capacities.

Third, as each copy of the ledger needs to be updated when new transaction(s) are added and the ledger changes, the greater the distribution of copies of the ledger the more time it will take, reaching the more physically far-flung nodes later (perhaps only seconds or parts of seconds, but nonetheless relevant in the context) than more centrally located nodes: hence, as well as taking account of this in design of the overall ledger, it may necessitate inclusion of a measure such as short periods (presumably only seconds, or even parts of seconds) when trading doesn't occur while the ledger updates, so that all market participants have the same accurate, up-to-date ledger information available at all trading times.

---

[2]   Each node being a computer address or location.

Finally, the greater the distribution of the ledger, the more careful ledger structural design will need to be to avoid 'choke points', where a hacking attack on the system, or technical breakdown, might isolate a portion of the network, thus causing that part of the ledger to be divergent from the rest (at least until the problem is identified and rectified).

Approaches to address these practical issues of market operation would be, primarily, a matter for technical design beyond the scope of this book. If required, it would be possible also to implement interfaces between existing[3] IT systems and the DL system. Thus, at the intra-jurisdictional level, if they so chose, entities could continue to operate their corporate systems on existing software and implement a software interface to the DL system, if the jurisdiction operated its ETS on a DL system. Similarly, if a jurisdiction wished to continue operating on a legacy system, it may be feasible to implement an interface between it and the DL network. The downside of this approach would be that the traceability, transparency, and immutability of the DL system would stop at the relevant interface, resulting in a sub-optimal network outcome overall. Hence, while from a technical perspective it is an option to accommodate retention of legacy software systems, while still implementing the DL at a higher level, the optimal outcome would be for the DL architecture to be fully diffused.

(ii) Additional infrastructure requirements
Apart from implementing the DL software if they choose to do so, existing ETSs would continue to operate on the same basis domestically, without any impediment or imposition, and the administrators would continue to perform their roles as before. Similarly, entities trading domestically in the ETS would continue to do so, without change. Participating jurisdictions would continue to be accountable for their respective ETSs, including the operation and integrity of their own registries, notwithstanding becoming part of the broader networked ledger, which would, in turn, were this to be done by implementing DLT domestically, facilitate their record keeping, accounting, audit, and reporting.[4]

(a) Network entity and standing management body
While participating jurisdictions would own, manage, and remain responsible for that part of the network operating inside their geographical area, at the network level there would be a need for a new entity to manage, operate, and

---

[3]   These are also described as 'legacy' systems.
[4]   See also: Justin D. Macinante, 'A Conceptual Model for Networking of Carbon Markets on Distributed Ledger Technology Architecture' [2017] *CCLR* 243, 250.

maintain the network to the extent that it would exist outside the individual jurisdictional boundaries. This 'network entity' could possibly be privately owned and operated, but it is considered more appropriate that it be owned by jurisdictions participating in the network, which would bear the costs of its provision and operation collectively. It would be accountable for the network overall, including in respect of reporting obligations to the Conference of Parties to the Convention, serving as the meeting of Parties to the Paris Agreement (CMA).[5]

Ongoing governance and operational management direction for the network entity could be provided by a standing management body. This body might be constituted at two levels, being an executive level, possibly in the form of a permanent secretariat, and a non-executive level made up of representatives of participating jurisdictions, with rotating presidency and membership, procedural rules, and schedule of periodic meetings.[6] The functions of the network entity would relate to the operation and maintenance of the physical and electronic components of the network and, in this sense, would appropriately be in the remit of the professional executive. The standing management body, in addition to overseeing and ensuring appropriate funding of the network entity to perform its functions, would monitor trading (international transfers), and ensure environmental integrity of the trading system, reporting, robust accounting, and audit across the trading network.

There is limited consideration in the literature given to governance of connected carbon markets and what there has been is, understandably, in the context of linking ETSs (see preceding chapter). For instance, a review of current research in 2015 even notes: 'governance of linking is an area of potential further exploration'.[7] In this context, it has been observed that arrangements may range from loose cooperation between jurisdictions through to an international organization with formal powers.[8] As that observation is about linking, the consideration given is in terms of the degree of integration or harmonization states may be willing to countenance under such institutional

---

[5]   Ibid. Considered in the next chapter.

[6]   There are examples of such supervisory structures in international and transnational organisations: for instance, the Bank for International Settlements: <www.bis .org>; the Financial Stability Board: <www.fsb.org>; International Organisation of Securities Commissions: <www.iosco.org>. The UNFCCC COP and Secretariat are another illustration of such a structure.

[7]   Aki Kachi et al., *Linking Emissions Trading Systems: A Summary of Current Research*, January 2015, ICAP, 11 <https://icapcarbonaction.com/en/?option=com _attach&task=download&id=241> accessed 06/09/17.

[8]   Michael Mehling, 'Linking of Emissions Trading Schemes' in David Freestone and Charlotte Streck (eds.), *Legal Aspects of Carbon Trading: Kyoto, Copenhagen and Beyond* (Oxford University Press, 2009) 122.

arrangements.[9] Issues identified to explore in that context include the legal form of the link, mechanisms for information, consultation, and conflict resolution.[10] The establishment of an international or supranational organization might occur as a later stage of integration and such an entity might have separate legal personality, a defined governance structure, and enforcement powers; power to collect market information, control over market access and accountability, oversight of market abuse, and management of prices.[11] The EU-Swiss ETS linking agreement,[12] for instance, establishes a Joint Committee composed of representatives of the parties[13] to administer the agreement and ensure its proper implementation. The Committee's functions are confined to discussing proposed amendments to the agreement, conducting reviews in the light of developments in either ETS and trying to settle any disputes that might arise.[14] Similarly, the California-Quebec agreement provides for a Consultation Committee.[15]

The elements described by Mehling would be applicable also to the network entity proposed here, which is envisaged performing a functional, operational role, ensuring the proper and efficient running of the networked market. It would have separate legal personality and its constitutional rules, to which jurisdictions would sign up, and the standard terms and conditions for transactions built into smart contracts, could address market governance issues. ETS administrators of jurisdictions participating would be part of the network and domestic financial regulators from those jurisdictions would continue to oversee the operation of markets domestically. It is proposed these domestic bodies – the ETS administrators and financial regulators – would also cooperate with (or their representatives would constitute) the standing management

---

[9]　Ibid.

[10]　Ibid, 123.

[11]　Ibid. Note also that another author, Joseph Aldy, has proposed a Bretton Woods type climate institution, however, this proposal corresponds more with the entity proposed by this book for overseeing mitigation value assessments (later in this chapter): see Joseph Aldy, 'Designing a Bretton Woods Institution to Address Climate Change', 2012, HKS Faculty Research Working Paper Series RWP12-017, John F. Kennedy School of Government, Harvard University <https://dash.harvard.edu/bitstream/handle/1/8830777/RWP12-017_Aldy.pdf> accessed 17/04/18.

[12]　Agreement between the European Union and the Swiss Confederation on the linking of their greenhouse gas emissions trading systems. OJ L 322, 7.12.2017, 3–26.

[13]　Ibid, Article 12.

[14]　Ibid, Article 13.

[15]　Article 12, Agreement between California Air Resources Board and the Government of Quebec, Concerning the Harmonisation and Integration of Cap-And-Trade Programs for Reducing Greenhouse Gas Emissions, 27 September 2013 <https://www.arb.ca.gov/cc/capandtrade/linkage/ca_quebec_linking_agreement_english.pdf> accessed 06/03/18.

body to ensure effective governance of the networked market, including maintenance of environmental integrity, and efficient operational management of, inter alia, matters such as price volatility, market abuse, market manipulation or dominant trader behaviour (see Chapters 8 and 9).

(b) Exchange, settlement platform and intermediaries

A further infrastructural consideration relates to the need for a trading exchange and settlement platform for transactions between jurisdictions. Whereas settlement for many securities transactions can take up to three days from the trade date (T+3), foreign exchange settlements two days (T+2) and US Treasury bonds even requiring one day (T+1),[16] one of the benefits of a decentralized infrastructure based on a network of nodes, in which participants are able to engage and transact with one another peer-to-peer (P2P) is that, by removing the need for trusted central counterparties to intermediate the transactions, there are various efficiencies '… by reducing the need for multiple intermediaries, the DLT could also reduce transaction costs'.[17] Reducing transaction costs improves the efficiency with which the market operates to achieve mitigation outcomes.[18] Thus, while the actual electronic transaction process may be marginally slower – seconds or even minutes, rather than microseconds – the end-to-end execution, settlement, and transaction completion theoretically should be faster overall.

The question is whether multiple levels of custody, clearing, and settlement might be reduced to a single platform, with trading taking place directly on the trading platform attached to the ledger, obviating the roles of broker, central clearing counterparty and clearing members, and, arguably, custodians (central securities depositaries and nominees).[19] While counterparty risk in cash 'spot' transactions (in the primary market) may be ameliorated by immediate settlement, counterparty risk extends throughout the life of a derivative (secondary

---

[16]   International Monetary Fund, 'Virtual Currencies and Beyond: Initial Considerations' January 2016, Staff Discussion Note SDN/16/03, 22.

[17]   European Securities and Markets Authority (ESMA), 'The Distributed Ledger Technology Applied to Securities Markets' Discussion Paper, 2 June 2016, ESMA/2016/773 9–13, at 12.

[18]   Transaction costs in relation to emissions market mechanisms and Coase theory discussed in: Navraj Ghaleigh, 'Two Stories about EU Climate Change Law and Policy' (2013) 14(1) *Theoretical Inquiries in Law* 43. As to the impact of administrative transaction costs in the EUETS: Peter Heindl, 'The Impact of Administrative Transaction Costs in the EU Emissions Trading System' (2017) 17:3 *Climate Policy* 314.

[19]   Stuart Davis and Julian Cunningham-Day, Linklaters LLP, 'Blockchain – recalibrating the market infrastructure', Going Digital Quarterly Breakfast Briefing, 14 October 2016, presentation Powerpoint slide 16.

market) contract, such as a future. Hence, given that derivatives will probably constitute a substantial part, if not the bulk, of the market,[20] central clearing counterparties are likely still to be required.[21] In relation to other intermediaries, whether these roles persist (and, consequently, the need for regulatory controls on them) will be a matter for how the conduct of transactions across the network is structured (see operational mechanisms, below).

Furthermore, settlement delivery against payment raises the issue of the need for an interface between smart contracts on the DL network and fiat money, for immediate settlement to be effective. In order that a DL networked market can deliver the time and cost savings of immediate settlement delivery against payment, there needs to be a way to provide for payment on the ledger. One possibility is to provide for payment through digital currencies, although this is not favoured as it would introduce another step (and related risks, for instance, volatility of digital currencies) to the settlement process.[22] An alternative is for an interface between the ledger and fiat currency system. This would build in an additional time factor, and also risk considerations, into the transaction process. All the same, it is noted that central banks and intergovernmental organizations are investigating both digital currencies[23] and the

---

[20]   Emissions derivatives constitute more than 90 per cent of the carbon market: European Court of Auditors, 'The integrity and implementation of the EU ETS' Special Report, 2015, paragraphs 12–24. <http://www.eca.europa.eu/Lists/ECADocuments/SR15_06/SR15_06_EN.pdf> accessed 23/06/17.

[21]   Fn.19 (Davis et al./Linklaters) slide 18.

[22]   It is noted that tokens or digital currencies tied to the value of the US$, other fiat currencies, or other classes of assets are being developed. However, these depend on the balance sheet and capital reserves of the entity or organization backing them, thus really need to be issued by central banks or substantial financial organizations to be considered viable. Some jurisdictions seem willing to accommodate a role for these so-called 'stablecoins', see: Swiss Financial Market Supervisory Authority FINMA, Supplement to the guidelines for enquiries regarding the regulatory framework for initial coin offerings (ICOs), 11 September 2019 <www.finma.ch> accessed 27/01/20; however, there are concerns over the risks they pose: Council of the European Union, Note from Presidency to Council: Stablecoins, Brussels, 6 November 2019, 13571/19, EF 314, ECOFIN 950 <https://data.consilium.europa.eu/doc/document/ST-13571-2019-INIT/en/pdf> accessed 27/01/20; Financial Stability Board, Regulatory issues of stablecoins, 18 October 2019 <https://www.fsb.org/wp-content/uploads/P181019.pdf> accessed 27/01/20; Bank for International Settlements, CPMI/IMF, G7 Working Group on Stablecoins, Investigating the impact of global stablecoins, October 2019 <https://www.bis.org/cpmi/publ/d187.pdf> accessed 27/01/20; European Central Bank, Occasional Paper Series, In search for stability in crypto-assets: are stablecoins the solution? No.230/August 2019 <https://www.ecb.europa.eu/pub/pdf/scpops/ecb.op230~d57946be3b.en.pdf> accessed 27/01/20.

[23]   Fn.16 (IMF); also see: Morten Bech and Rodney Garratt, 'Central Bank Cryptocurrencies' (September 2017) *BIS Quarterly Review* 55–70 <https://www

possibility of adapting central bank settlement systems to facilitate emerging settlement and payment infrastructures to access central bank money,[24] thus potentially affording a solution.

2.   Operational Rules of the Distributed Ledger Network[25]

Jurisdictions that join the network would need to abide by certain operational rules necessary for the functioning of the network. As proposed here, these rules would be part of the contractual terms by which jurisdictions subscribe to the constitution and become part owners of the network entity. They would include arrangements as to permissioning, consensus, and encryption.

Private/public key encryption is one of the elements that identifies a DL system, while permissioning is a configurable element that defines both who is able to join the system and their level of access to, and interoperability with, the ledger – notwithstanding that they may hold the ledger in full on their computer. In the model proposed, the ledger may be organized as a chain of blocks of information – each block containing a collection of transactions. Transaction information is, thus, exchanged between nodes and added as a new

---

.bis.org/publ/qtrpdf/r_qt1709.pdf> accessed 24/01/18; Robleh Ali, John Barrdear and Roger Clews, 'Innovations in Payment Technologies and the Emergence of Digital Currencies', Bank of England Quarterly Bulletin, 2014 Q3 <http://www .bankofengland.co.uk/publications/Documents/quarterlybulletin/2014/qb14q3d igitalcurrenciesbitcoin1.pdf> accessed 12/01/17; John Barrdear and Michael Kumhof, 'The Macroeconomics of Central Bank Issued Digital Currencies', July 2016, Bank of England, Staff Working Paper No.605 <http://www.bankofengland.co.uk/research/ Documents/workingpapers/2016/swp605.pdf> accessed 12/01/17.

[24]   In 2018, the Bank of England announced it was conducting Proof-of-Concept exercises with technology companies to understand how its Real-Time Gross Settlement (RTGS) service could support settlement in systems operating on innovative payment technologies, see: Bank of England, 'RTGS Renewal Programme Proof of Concept: Supporting DLT Settlement Models' <https://www.bankofengland.co.uk/-/media/boe/ files/payments/rtgs-renewal-programme-proof-of-concept-supporting-dlt-settlement -models.pdf?la=en&hash=894DFF2C6DE88434EA3A96612E9FD6F454BF68DA> accessed 06/08/18; while the Bank of Canada, Bank of England, Bank of Japan, European Central Bank, Sveriges Riksbank and the Swiss National Bank, together with the Bank of International Settlements, announced on 21 January 2020 forma- tion of a group to 'share experience as they assess the potential cases for central bank digital currency (CBDC) in their home jurisdictions.' News Release, Bank of England, 21 January 2020 <https://www.bankofengland.co.uk/-/media/boe/files/news/ 2020/january/central-bank-group-to-assess-potential-cases-for-central-bank-digital -currencies.pdf> accessed 31/01/20. Note also the discussion following in Chapter 9 in relation to: European Securities and Markets Authority (ESMA), 'Advice on Initial Coin Offerings and Crypto-assets', 2019, ESMA50-1391.

[25]   This sub-section builds on the author's prior publication: fn.4 (Macinante).

ledger entry to the computers of all participants (nodes). In the absence of a trusted central party, the updating in this way relies on a consensual process amongst the nodes.

(i)   Consensus

To briefly recap, distributed consensus involves two steps: first, there is validation of each transaction, which involves certain nodes (validating nodes) checking every transaction in the block of transactions (assuming the network is blockchain based) to ensure that each transaction is legitimate. For example, they would verify that the sender of a transaction owns, or can otherwise deal with, the asset being sold. The second step is broadcast and consensus, which takes place after a validating node has validated one or more transactions: it then initiates addition of the transaction data to the ledger by broadcasting information about this new block or entry to other validating nodes, who collectively resolve (through their pre-agreed consensus mechanism) a common set of validated transactions to be added to ledger.[26]

(ii)   Permissioning

Permissioning signifies, first, that the network is not open. In other words, it is set up so that only trusted or vetted participants can participate in the control and maintenance of the system. The network proposed here is a permissioned network, meaning only authorized entities can participate. Thus, in order to trade on the network, an authorization to trade in the domestic ETS of a jurisdiction that joins the network needs to be held.[27]

The degree of access an entity has to the ledger would depend on the nature of its activity. For instance, entities with compliance obligations under the ETS (compliance entities), other authorized traders, and market makers (not subject to compliance obligations) would, perhaps, be able to access that part of the ledger relating to their own accounts and transactions (possibly even just their current transactions on the ledger, but not their full history of transactions). On the other hand, the administrator of the ETS would have access to all accounts and transactions pertaining to its own registry, but not necessarily have access to the entire distributed ledger. Each jurisdiction's ETS registry administrator might be authorized to see also the composite ledger records for each of the other participant jurisdictions, but certain restrictions might be regarded as necessary, for example: in regard to the entries relating to individual entities (of other jurisdictions), in particular because of data protection and confidentiality concerns; and possibly even regarding access to the ledger as a whole,

---

[26]   Ibid, 260.
[27]   Ibid, 254.

to information which, for a variety of reasons, may be considered nationally sensitive (for example, relating to or considered indicative of the strength of a national economy).[28] Public access to information might be through the ETS administrator in each jurisdiction.

(iii)   Private/public key encryption

Private/public key encryption is an example of asymmetric key encryption, using a pair of keys with the public key used to encrypt the data and the private key being held by the recipient who is to decrypt it. This permits strangers to exchange confidential data on a public network, without concerns over security and without sharing a single key.

As mentioned earlier, the robustness of DLT in protecting the integrity of information is due, at the macro-level, to design including, for instance, the consensus process amongst validating nodes. At the micro-level, it is due to the cryptographic technology applied. The level of transparency and replication of records provides a certain buffer against fraud, hacking, and other possible abuse or corruption.[29] It would be incumbent on the jurisdictions joining and subscribing to the network to ensure that the entities they authorize to trade on the DL network put in place sufficient security measures to protect private keys. Any losses resulting from failure by an authorized entity to protect its private key should, logically, fall on that entity.

3.   The Operational Mechanisms Required

Before considering possible rules governing how market transactions might proceed, it is necessary to outline two mechanisms that, while both essential to the operation of the market proposed, pertain to differing aspects. The first mechanism relates directly to climate policy, namely, how differences between the efforts of different jurisdictions in terms of mitigation might be valued and how this value possibly translates into a conversion factor for mitigation outcomes transferred between jurisdictions, and so, a price differential. The second relates to how the transaction proceeds, so in one sense is purely mechanistic, yet in another sense, also relates back to climate policy.

---

[28]   Ibid. Although, such a restriction may run counter to the objective of greater transparency in the Paris Agreement.

[29]   ASTRI, 'Whitepaper On Distributed Ledger Technology', (11 November 2016) Commissioned by Hong Kong Monetary Authority <http://www.hkma.gov.hk/media/eng/doc/key-functions/finanical-infrastructure/Whitepaper_On_Distributed_Ledger_Technology.pdf> accessed 12/1/17.

(i)   A mechanism for valuing mitigation[30]

Chapter 5 introduced the concept of networking as a way of connecting markets, based on a model that values the traded emission allowance unit to take account of that unit's actual worth in terms of mitigation achieved. This recognizes the existing global diversity of mechanisms for carbon pricing as such mechanisms, although reflecting local preferences, are fragmented and heterogeneous. As a result, differences in design, implementation, and standards detract from their effectiveness.[31] It takes account also of the heterogeneity of approaches recognized by the Paris Agreement.

Networking carbon markets[32] identifies as one of its three core elements a mechanism to measure value of mitigation outcomes: 'an independent assessment framework to guide and assess the implementation of climate actions'.[33] Thus, networking is based on the notion that a measurable value, the mitigation value (MV) can be placed on the outcomes generated by an ETS.[34] This section introduces the concept and briefly addresses the mechanism by which such an assessment can be made.

There are differing understandings as to what the expression 'mitigation value' entails[35] and, more so, how the MV of a mitigation outcome is determined. Early reports of the negotiations concerning guidance on cooperative approaches referred to in Article 6, paragraph 2 of the Paris Agreement, indicated that discussions touched on related issues, for example, in relation to environmental integrity, the quality of units.[36] Yet, considering its relevance to the negotiations, the lack of discussion concerning mitigation value is disheartening, while there seems to have been only passing consideration of the valuing of emission trading scheme units, in negotiations or in the literature.[37]

---

[30]   This sub-section builds on the author's publication: Justin D. Macinante, 'Operationalizing Cooperative Approaches Under the Paris Agreement by Valuing Mitigation Outcomes' [2018] *CCLR* 258.

[31]   Fn.4 (Macinante) 244.

[32]   See Chapter 6 supra as to source of concept.

[33]   World Bank Group, Networked Carbon Markets: Mitigation Action Assessment Protocol, 2016, World Bank, Washington, DC. © World Bank, 8 <https://openknowledge.worldbank.org/bitstream/handle/10986/25371/110153-WP-P161139-PUBLIC-MAAPMay.pdf?sequence=1&isAllowed=y> accessed 27/02/18.

[34]   Fn.4 (Macinante) 245.

[35]   See, for instance, World Bank's Networked Carbon Markets webpage, documents on concept development <http://www.worldbank.org/en/topic/climatechange/brief/globally-networked-carbon-markets>.

[36]   Informal note by the co-chairs, Third iteration, 12 November 2017, Subsidiary Body for Scientific and Technological Advice, Forty-Seventh meeting, <http://unfccc.int/files/meetings/bonn_nov_2017/in-session/application/pdf/sbsta47_11a_third_informal_note_.pdf> accessed 27/02/18.

[37]   Fn.30 (Macinante) 262.

To date, the majority of studies on connecting schemes seem to have focused on full bilateral linking under which the units are fully fungible in all participating systems,[38] which may partially explain this absence of consideration. The fact of the homogeneous approach taken prior to the Paris Agreement, that is, under the Kyoto Protocol,[39] where the value of all traded units was defined as being equal to one tonne CO_2-equivalent GHG,[40] may be another reason.

All the same, there is some discussion of MV to be found. For instance, Aldy noted that assessments of mitigation value could play an important role in linking between countries with disparate mitigation policies.[41] These MV assessments, it was speculated, could inform the linking agreement through exchange rates which, if transparent, could be used to incentivize higher ambition on the part of more poorly performing jurisdictions.[42] Another author, Mehling, observes that while a move from a regime based on compatibility of systems and equivalence of traded units, to one that seeks to quantify and compare mitigation effort, offers interesting perspectives, it will also give rise to political controversy and raise similar challenges to those experienced in negotiations to date.[43] Two responses are briefly mentioned here.[44]

First, it is argued by Aldy that the application of analytical tools for data gathering 'is crucial for assessing the country-level, comparative, and aggregate impacts of those efforts'[45] and these tools and associated data 'in turn rely on effective transparency and review mechanisms'.[46] Aldy demonstrates that the idea of transparency and policy surveillance of countries in the context

---

[38]   Fn.7 (Kachi et al./ICAP) 10.

[39]   Discussed in Chapter 4 supra.

[40]   Fn.30 (Macinante) 262.

[41]   Joseph E. Aldy, 'Evaluating Mitigation Effort: Tools and Institutions for Assessing Nationally Determined Contributions' Cambridge, Mass.: Harvard Project on Climate Agreements, November 2015 <http://pubdocs.worldbank.org/en/736371454449389076/pdf/Evaluating-Mitigation-Effort-Nov-2015.pdf> accessed 27/02/18. The work of Aldy and his colleagues in this respect has been reviewed in the author's previous publication: see fn.30 (Macinante) 262.

[42]   Ibid (Aldy) 32. See also: Michael Lazarus et al. (Stockholm Environment Institute), *Options and Issues for Restricted Linking of Emissions Trading Systems*, September 2015, ICAP Berlin, Germany <https://icapcarbonaction.com/en/?option=com_attach&task=download&id=279> accessed 06/09/16.

[43]   Michael Mehling, 'Legal Frameworks for Linking National Emissions Trading Schemes' in C. Carlarne, K. Gray, and R. Tarasofsky (eds.), *Oxford Handbook of International Climate Change Law* (Oxford University Press, 2016) 276.

[44]   These are set out in the author's previous publication at fn.30 (Macinante), but given the centrality of the concept to the model proposed, are summarized again for completeness.

[45]   Fn.41 (Aldy) 13.

[46]   Ibid.

of multilateral regimes, is not something new,[47] referring to a number of transparency models from other multilateral regimes, such as the International Monetary Fund (IMF) annual country-level economic surveillance; the Organisation for Economic Cooperation and Development (OECD) peer reviews of member states' economic policies every one or two years; and the World Trade Organization (WTO) regular reviews of members' trade policies. The conclusion is that there is an array of models to which the international community can resort.[48]

Furthermore, countries participate in sovereign credit rating assessments in order to borrow. For example, Standard & Poor's sovereign issuer credit ratings evaluate a sovereign's ability and willingness to service financial obligations to commercial creditors. They comprise a framework including policymaking; income levels, GDP per capita, tax and funding bases; currency in international transactions, external liquidity, residents' assets and liabilities relative to rest of the world; sustainability of debt burden; and exchange rate regime and monetary policy credibility.[49] Countries' concerns about such statistics do not arise when the objective is access to international debt markets. This book argues that there should not be a difference when it comes to accessing an international carbon market, from which there may be similar economic benefits to be reaped.

The second response is that many of the sources of potential political controversy can be addressed through careful design. For instance, it is argued that elements of the regime proposed here ameliorate the causes of potential political controversy:

- by ensuring the independence of the process to quantify and compare mitigation effort and that the entity or entities carrying out that assessment comprise relevantly qualified, independent, impartial experts;
- by applying generic criteria to assessments uniformly across all jurisdictions in that process, such that all jurisdictions are subject to equivalent treatment under the process;
- ensuring that the process and outcome are open and transparent and that outcomes are communicated appropriately as market sensitive information; and
- affording all jurisdictions the flexibility to engage with, or leave, the process relatively easily and on the same basis – in the event that, as an

---

[47]   Ibid.

[48]   Ibid, 34.

[49]   S&P Global, 'Sovereign Rating Methodology', 2017 <https://www.spratings .com/documents/20184/4432051/Sovereign+Rating+Methodology/5f8c852c-108d -46d2-add1-4c20c3304725> accessed 01/03/18.

information tool, the assessment is part of an agreed governance framework (as opposed to being purely private sector driven).[50]

It is beyond the scope of this book to postulate a specific methodology for determining the MV of mitigation outcomes. Rather, a couple of possible approaches to how a methodology might be applied are suggested. The first of these possible approaches would be for an organization akin to the Clean Development Mechanism Executive Board (CDMEB) model. The second would be for MV assessments by private sector entities, perhaps similar to credit reference agencies (CRAs), and subject to regulation similar to that now administered (in relation to CRAs in the EU) by the European Securities and Markets Authority (ESMA).[51] The details of the latter, private sector approach are set out in an earlier publication.[52] In short, it would entail private sector (CRA-type) entities being accredited to assess and determine MVs, based on approved methodologies, subject to authorization and supervision along the lines of the ESMA regulatory model. The outcomes would be publicly available market information.

Given aspects of the CDMEB experience, the first model may be more problematic in fostering re-engagement in the market by the private sector.[53] Some of the issues included absence of transparency, clarity, and predictability in decision-making; and absence of decision review or appeal rights.[54] Its make up as a regional negotiating group nominees, rather than a panel of independently assessed, expert appointees, has been noted.[55] The CDM process has been lengthy and cumbersome.[56] The CDMEB's role, which includes being de facto gatekeeper over project flow, has been problematic as well, pointing to the need for any MV process to separate the function of regulating providers of MV, from the actual provision of MV, which should just be market information, available independently of market operation.[57]

---

[50] Fn.30 (Macinante) 263.

[51] Ibid, 269–70.

[52] Justin Macinante, 'Networking Carbon Markets – Key Elements of the Process', 2016, World Bank Group Climate Change, 33–40 <http://pubdocs.worldbank.org/en/424831476453674939/1700504-Networking-Carbon-Markets-Web.pdf> accessed 01/03/18.

[53] Dependence of this market on investor confidence has been flagged by: Charlotte Streck and Jolene Lin, 'Making Markets Work: A Review of CDM Performance and the Need for Reform' (2008) 19(2) *European Journal of International Law* 409, 420.

[54] Fn.30 (Macinante) 269.

[55] Ibid. Also see: Ilona Millar and Martijn Wilder, 'Enhanced Governance and Dispute Resolution for the CDM' [2009] *CCLR* 45. Recommendations for how the issues can be rectified are noted, as are the alternative models raised by these authors.

[56] Fn.53 (Streck and Lin).

[57] Fn.30 (Macinante) 269.

Notwithstanding these observations, as noted, proposing a process for delivering an MV methodology is beyond the scope here. However, the regulatory and institutional frameworks within which the model proposed might operate are the focus here so, to this extent, the body or entities that might carry out the MV assessments and the regulatory and institutional structures that may exist for that purpose are relevant. Therefore, in order to carry out that analysis (in the next chapter) it is proposed to proceed on the basis of the private sector model, since this is perceived more likely to engage private sector involvement.

(ii)   The mechanism required for effecting transactions[58]

There are different possible mechanisms by which transactions could proceed. For instance, a jurisdiction might cancel the units being transferred at the time of transaction and, the buyer having purchased them, the receiving jurisdiction create and credit them to the buyer's account in its registry. This assumes that the units from the respective jurisdictions would be fully fungible. However, such a process could be administratively cumbersome, involving coordination of registries across jurisdictions.

An alternative to the movement of emission units from one jurisdiction to another, for instance where they are not fully fungible, would be to have a 'transaction unit' (TU).[59] The transaction mechanics would be for the transferring jurisdiction to convert its units into TUs,[60] following which, the buyer having purchased them, the receiving jurisdiction would convert the TUs into its domestic ETS units.[61] In this instance, the TU serves as the medium of exchange.

Using TUs as a medium of exchange might be considered useful for a number of reasons, including reducing the number of conversion rates needed in a multilateral system; the benefit of fewer, but larger and more liquid markets with fewer asset (unit) balances; reducing the information required by participants, fostering simple and cheap operations, with more efficient transactions; less opportunity for improper market behaviour such as fraud; and an overall reduction in administration and transaction costs.[62] All the same, it is not necessary to propose a conclusive basis upon which TUs might be founded, but simply to note that interposing a transaction unit mechanism may facilitate

---

[58]   This section draws on author's previous publication: fn.4 (Macinante).

[59]   Ibid, 247. The role of such a transaction unit would be analogous to the role of the US$ in international currency transactions.

[60]   Ibid. The transferring jurisdiction's units would be cancelled when the transaction units are created.

[61]   Ibid. The transaction units would be cancelled when the receiving jurisdiction's units are created and credited in its registry.

[62]   Ibid, 248.

the transactional process, especially once the distributed network goes beyond two participant jurisdictions.[63]

4.   Transactional Rules as Part of the Regulatory Framework[64]

Market participants would access the DL by virtue of being authorized to trade in a participant jurisdiction's domestic scheme. Each participant jurisdiction would continue to maintain its registry for that purpose and impose rules on entities from its own ETS that are authorized to use the network. For inter-jurisdictional emissions trading that conforms to international (and national) climate change policies, certain rules and principles would be fundamental. For instance, the rules that govern the relationships between jurisdictions in the networked market (that is, the rules the jurisdictions commit to uphold when joining the network) would themselves, to a degree, be transposed into the code for the terms and conditions of contracts between counterparties to transactions.[65] Thus, these rules would operate on two distinct levels, first, in the form of terms and conditions to which a jurisdiction would need to subscribe in order to join the network entity; and second, as part of the standard terms and conditions applicable to each individual transaction.

Illustrations of possible standard terms and conditions governing transactions that would be applied automatically through the code of smart contracts are set out, by way of illustration, in the Appendix. Transactions on the network would apply the standard terms and conditions embodied in electronic code (smart contracts). A number of the rules mentioned above would be transposed into these standard contract terms and conditions, particularly for example, those in (i)(a)–(e) in the Appendix. Thus, the commitments given by

---

[63]   Issues related to this element are relevant to and considered as part of the analysis of regulatory and institutional frameworks in Chapters 8 and 9.

[64]   Section draws on the author's previous publication: fn.4 (Macinante). The regulatory framework is analyzed in the next chapter, however, the transactional rules straddle both the elements of the market proposed (hence mentioned here) and the regulatory framework within which it would operate.

[65]   It is noted that participating jurisdictions would, at all times, retain jurisdictional control over the entities authorized by them to trade – including trading in the networked market. Hence, in the circumstances of incompatibility between the rules of a participating jurisdiction and the trading rules of the networked market, the entity would be obliged to follow the rules of its domestic jurisdiction. However, given the fact that jurisdictions, when opting to join, can decide the basis on which they authorize their entities to participate (e.g. any limits or boundaries that apply), the likelihood of such a situation of incompatibility arising between rules of a jurisdiction and NCM trading rules is considered remote.

the jurisdictions in joining the network would be automatically applied by the entities they authorize in the transactions undertaken.

5.   Participants at Different Jurisdictional Levels

(i)   Jurisdictional

The ETS administrators of participating jurisdictions would be non-trading actors in the network. The registries administered by them would themselves be part of the distributed ledger, to which they would have access for monitoring and verification of transactions involving counterparties authorized by them, and for examining data for audit and reporting purposes. A range of entities under domestic ETSs would be authorized, including those that have compliance obligations (compliance entities) and those that are participating for other commercial reasons as traders, brokers, and market makers (for example, on behalf of clients), and do not have specific compliance obligations under the ETS.[66] The computer code for contracts would need to distinguish entities that were subject to compliance obligations under a jurisdiction's ETS, from those that were not.[67] The absence of specific compliance obligations on entities would mean that certain conditions, such as compliance reserve obligations (see Appendix), would not apply to them individually (although they might still be applicable on a jurisdictional basis). The domestic financial regulator in participating jurisdictions would also have a role, both regulating the activities of market participants in the jurisdiction and by contributing to the broader governance process through participation in a supra-jurisdictional financial supervisory/advisory body (see following chapter).

(ii)   Cross-jurisdictional

As noted earlier in relation to additional infrastructure requirements, participating jurisdictions could establish a standing management body to ensure ongoing governance and operational management of the network entity operating the distributed ledger platform. There are a number of models in existing international bodies for how this might be structured and conduct its business. For example, as noted earlier, there might be a standing secretariat that would call together jurisdictional representatives for either regular periodic, or ad hoc meetings. This would also necessitate procedural rules.

These elements would generate additional financing requirements for staff, equipment and premises, legal drafting and advice, financial management and accounting, and so on. Nevertheless, it is expected that such additional costs

---

[66]   Fn.4 (Macinante) 255.
[67]   Ibid.

(shared by the participating jurisdictions perhaps according to a formula, for instance, based on the volume traded or another metric) would be minimal in comparison with the value the distributed market could achieve. The addition of a standing management body may avoid the need for supervisory or controlling mechanisms in jurisdictions' own administrations, thereby lowering costs overall in the longer term.[68] Even if transactions in the primary market were to be peer-to-peer, without the need for intermediaries such as central counterparties, there would need to be an exchange or trading platform where transactions take place. This would be part of the ledger function and, as such, come under the purview of the network entity.

Additionally, as flagged earlier and based on experience in the EUETS, transactions in the derivative (futures) market are likely to make up the bulk of trading. Since counterparties' positions can remain open in futures contracts for the term of the contract, it is likely that there will be a need for provisions to reduce structural market risk due to defaults. Thus, at least in so far as futures contracts trading is concerned, there may be the need for centralized clearing. All the same, for the purpose of this discussion consideration is confined to the primary market as this has most direct relevance to the actual transfer of mitigation outcomes between entities. Assuming that MV assessment is carried out by private sector entities, possibly under a model similar to that which operates for CRAs, those entities could also be seen as operating at a cross-jurisdictional level since, it is envisaged, the aim would be that they not be associated with any particular jurisdiction.

(iii)   Supra-jurisdictional
Two overriding supervisory bodies, acting conjointly, are proposed to take account of the fact that, first, the model, transactions, and market proposed are designed principally to give effect to the international transfer of mitigation outcomes (as understood in the Paris Agreement): thus, ultimately one of the overriding supervisory bodies should be a subordinate body of the CMA; and second, the nature of the proposed networked market as a financial market means that the other body would need to be a financial supervisory body. The nature and roles of these bodies and the regulatory framework within which they might operate are considered in the following chapter.

Finally, for the regulation of private sector entities undertaking MV assessments, there would be a need, at the supra-jurisdictional level, for a regulatory body, possibly along similar lines to the European Securities and Markets Authority (ESMA) and directly answerable to the overriding supervisory bodies, acting conjointly. The role of this body would include licensing MV

---

[68]   Ibid, 256.

assessors, supervising the assessments, and approving and certifying the methodologies applied. This role and these functions are also analyzed as part of the institutional and regulatory framework for the proposed market, in the following chapter.

ALTERNATIVES FOR IMPLEMENTING ARTICLE 6 OF THE PARIS AGREEMENT

Before moving to examine potential institutional and regulatory frameworks for the market model proposed here, possible alternative approaches to implement Article 6 of the Paris Agreement should be considered. As noted, the proposal here consists of two elements, thus alternatives might be considered in terms of those not involving networking; those that do not apply distributed ledger technology; and those applying neither element. Thus, possible alternatives might include:

- first, international trading of mitigation outcomes between the ETSs of individual, unconnected jurisdictions, either (a) via a globally centralized registry and transaction log, or (b) on a distributed ledger platform;
- second, trading taking place in clusters of linked jurisdictions forming homogeneous 'club' structures, either (a) via a globally centralized registry and transaction log, or (b) on club-based distributed ledger platforms, but without trading taking place from one such club to another; or
- third, international trading of mitigation outcomes in a networked market on a globally centralized registry and transaction log, with a CDM Executive Board-type body policing compliance (the ITL model).

Reasons have been set out[69] why connecting jurisdictions is considered more appropriate and beneficial than jurisdictions remaining unconnected. These reasons are valid in the case where unconnected jurisdictions trade with each other. A primary issue for such trading would be how the units traded would be valued and accounted for, raising a further question of whether this approach might also require inclusion of mitigation value assessments to ensure fungibility. Notwithstanding that the jurisdictions under this option are not connected by linking or networking, the need for a mechanism by which the values of units traded are derived points to the need for an agreement or treaty. Thus, any benefit from not negotiating a treaty to connect, would be cancelled out by the need to negotiate an agreement on how to value respective units. All the same, for the reasons outlined earlier, this book contends that connecting ETSs offers greater benefits.

---

[69] See Chapter 5.

In relation to the second alternative, reasons have been advanced also why networking is favoured over linking.[70] These apply irrespective of whether the club structures would have their own DL platforms or there would be a globally centralized registry and transaction log. It is acknowledged that there would be benefits to be gained by operating the club structures on DL platforms; however, realization of these benefits would still be subject to successful negotiation of the linking treaty. The fact that the jurisdictions were linking would remove the need for the development and implementation of a process to assess mitigation value, since the linking process would presumably include determination of a basis for equivalence, or at least alignment, of the carbon units/assets of the respective linking jurisdictions. All the same, negotiation of the terms for alignment of jurisdictions' schemes may be both time-consuming and difficult.

With respect the third alternative, a networked carbon market could operate on the ITL model. In a sense, this may be an easier alternative, as the ITL structure exists under the Kyoto Protocol and could be adapted to accommodate trading on a networked basis. On the other hand, the work required to carry out that adaptation may prove to be substantial, noting that the ITL deals with country-to-country transactions, thus would need to be able to accommodate transactions between authorized entities. More significantly, perhaps, would be whether and how the ITL could adapt from the existing binary checking function it performs to providing a more substantive ledger function, including an exchange platform with settlement and clearing. This would necessitate building in a mechanism to measure the mitigation value of mitigation outcomes: an independent assessment framework to assess the implementation of climate actions. Negotiation of a treaty for such would be necessary, unless the private sector were to step in and drive development of such a mechanism, thereby facilitating adoption by jurisdictions participating in trading.

While alternative approaches such as one including a centralized ledger like the ITL certainly are possible, this book contends that to continue with a centralized model would be an opportunity missed, given the technological developments taking place that have particular application to how both government services and financial sector services may be delivered in the future. Rather than moving forward in lock-step with exploration of these developments by the financial sector, re-engagement of which in the carbon market is important to the success of carbon pricing as a mitigation mechanism, continuing with a centralized model may simply result in re-compartmentalization, detracting from the effectiveness that carbon pricing can have in changing behaviour. Recognition of the heterogeneity of approaches to mitigation, including

---

[70] See Chapter 6.

cooperative approaches involving the international transfer of mitigation outcomes in the Paris Agreement has enabled these favourable circumstances; the networking of carbon markets can provide a mechanism; and the decentralized nature of distributed ledger technology is consistent with both the heterogeneity of approaches to mitigation and the disaggregated nature of networking. Together they might provide the facilitative platform on which the effectiveness of carbon pricing can be maximized, as this book extols.

PART IV

Analysis of the proposal

8. Governance structure for the networked market

Having proposed a model for networking carbon markets, it is important that consideration be given to the governance structures within which such a networking of markets might operate. Accordingly, this chapter examines the regulatory and institutional frameworks and relationships that comprise the governance structure for the proposed networked market. Governance, as used here, is taken to be the process through which state and non-state actors interact to design and implement policies within a given set of formal and informal rules that shape and are shaped by power.[1] Thus, the expression 'governance structure' is used here in a very broad sense to include the regulatory and institutional frameworks, involving both state and non-state actors and both formal and informal rules, established to implement greenhouse gas (GHG) mitigation policies and, specifically, emissions trading.

Analysis aims, first, to determine how well the governance structure accounts for the requirements of the three areas of law in which it must function, namely climate change law; financial markets regulation; and the legal requirements developing in relation to distributed ledger technology (DLT) and its applications. Each requires a different approach as, for example, in relation to climate change law, there is an existing international governance structure with which comparative analysis can be made; financial markets regulation is, on the other hand, principally a matter for domestic law, although there is a developing global structure which can be considered in terms of how the governance structure proposed here could fit; while DLT and its applications are new, so jurisdictions are currently active in formulating approaches, thus the approach taken examines these developments and assesses compatibility. Second, the analysis focuses specifically on the regulatory frameworks for these areas of law, their point of intersection and the particular issues to which they give rise for the governance structure.

This chapter begins by dissecting the governance structure vertically into three pillars, demonstrating the differing functionality of each; and then horizontally, to show seven tiers of governance. The structure is compared with

---

[1] World Bank, *World Development Report 2017: Governance and the Law* (Washington, DC, 2017) 3.

the existing carbon market governance structure for international emissions trading (IET) that developed under the Kyoto Protocol. How the proposed governance structure could fit into global financial market governance is considered in the third section, reviewing the institutions involved, their structures and roles. The final section then focuses on the responses to the advent of distributed ledger technology (DLT) applications in financial markets, in regulatory and analytical terms, interrogating those responses for how they might inform the application to networking of carbon markets.

The chapter following (Chapter 9) continues analysis in this framework with an examination of regulatory issues arising in relation to the proposed market, whether those issues pose particular difficulty for the governance structure envisaged and consideration of how those issues might be addressed. This analysis aims to demonstrate, so far as possible in terms of the hypothetical market proposed, first, that the governance structure set out is suitable for emissions trading in the context of the Paris Agreement; and second, that allowing the networked market to operate as a global financial market, with as little intervention as possible so as to maximize efficiency and effectiveness, but within a well-designed boundary framework of climate change rules, can promote the objectives of climate policy.

THE GOVERNANCE STRUCTURE

1.   Three Pillars of Functionality

Vertical dissection of the governance structure for the proposed networked carbon market shows it consisting of three pillars, each with differing functionality. The three, interacting pillars of functionality are: first, supervisory/regulatory; second, self-regulatory market operation; and third, the provision of market information, as set out in the following Table 8.1.

(i) Supervisory/regulatory
There are two aspects to this first pillar. First, is the dual nature of the institutional framework within which the market will operate: if not apparent already from the nature of the market itself, this duality is borne out by the proposal that there be two overriding supervisory bodies – one from the climate policy perspective and one from the financial market perspective – acting conjointly to ensure: (a) that the operation of the market is efficient and does not impact global financial stability in any way; and (b) that the market operates effectively in promoting and enhancing global mitigation efforts towards achieving the higher ambition envisaged by the Paris Agreement.

The second aspect to this pillar is the involvement of domestic financial regulators in jurisdictions participating in the networked market. While it is

Table 8.1   Vertical analysis: three functional pillars interacting

Supervisory/ Regulatory		Self-regulatory Market Operation		Independent Market Information
Overriding Supervisory Bodies		**Smart Contracts** Rules as code	←	MV assessment
Climate UNFCCC (COP/CMA/subsidiary body)	→	* Transaction rules * Environmental boundaries		Regulatory body (ESMA model) ↓
Global Financial (IOSCO committee)	→	**Network of carbon markets**	←	Assessors (CRA model)
* acting conjointly		Administrators of participating ETSs as permissioned owners		* independent, objective, technical process * regulatory body
* domestic financial regulators of participating jurisdictions	→	Authorised entities trading		independent but overseen by and ultimately answerable to Supervisory Bodies * assessors from the private sector

not essential that, in line with the approach taken by the EU, jurisdictions that have implemented domestic emission trading schemes (ETSs) legislate to treat emission units traded in those schemes as financial instruments, it would be desirable and ultimately may become the default position for jurisdictions wishing to join the networked market. It would afford greater consistency if they were to do so, as well as establishing the basis for domestic financial regulators from those jurisdictions to participate in and contribute to the work of the overriding financial supervisory body.

In relation to the conjointly acting supervisory bodies, it is noted that the Conference of Parties (COP) remains the supreme policymaking body, decision-making body and negotiating forum, of the UNFCCC and, acting as the Meeting of Parties to the Paris Agreement (CMA), does so for the Paris Agreement.[2] Thus the ultimate function of the supervisory bodies acting conjointly would be to advise, inform and report to the CMA.

The CMA is to keep implementation of the Paris Agreement under review and, within its mandate, make decisions to promote effective implementation, including establishment of such subsidiary bodies as deemed necessary for implementation and exercise of such other functions as may be required.[3] This

---

[2]   Article 16, paragraph 1, Paris Agreement.
[3]   Article 16, paragraph 4, Paris Agreement.

mandate includes supervision of international transfers of mitigation outcomes by Parties engaged in cooperative approaches under Article 6 and this could be undertaken through a climate subsidiary body established for that purpose (which would act conjointly with the financial supervisory body established under the auspices of a financial intergovernmental body or organization).

The Clean Development Mechanism Executive Board (CDMEB) provides a model for such a subsidiary body.[4] It is small (ten members), with members nominated from regional groupings,[5] although they serve for limited periods, must possess appropriate technical and/or policy expertise and act in their personal capacity.[6] Thus, the climate supervisory body might have a membership based on the number of jurisdictions participating in the networked market, with members required to possess technical and/or policy expertise and to act in their personal capacity.

With respect to the financial supervisory body, the role of providing supervision over this new inter-jurisdictional financial market might be allocated to an existing body, or a new committee or subordinate body of an existing body. There are a number of possible, relevant existing organizations in this respect, such as the Bank of International Settlements (BIS),[7] Financial Stability Board (FSB),[8] International Organization of Securities Commissions (IOSCO),[9] or the Financial Action Task Force (FATF).[10] The nature of these organizations and their roles suggests the function of providing supervision for an inter-jurisdictional carbon market might potentially come within the remit of any of them. All the same, bearing in mind the role envisaged (under this pillar) for domestic financial regulators of participating jurisdictions, a committee under the auspices of IOSCO would appear to be most appropriate, given IOSCO's role in relation to securities markets and also the range of committees carrying out current policy work under the backing of its Board.[11] Membership of such a committee could be constituted by representatives of financial regulators from jurisdictions participating in the proposed market.

---

[4]   Although note that it is constituted specifically in Article 12, paragraph 4 of the Kyoto Protocol, as opposed to under Article 13, paragraph 4(h), which corresponds to Article 16, paragraph 4(a) Paris Agreement.

[5]   UNFCCC COP7: Report of the Conference of the Parties on its Seventh Session, held at Marrakesh from 29 October to 10 November 2001, FCCC/CP/2001/13/Add.2, 21 January 2002, Decision 17/CP.7, annex, paragraph 7 <https://unfccc.int/sites/default/files/resource/docs/cop7/13a02.pdf> accessed 22/10/18.

[6]   Ibid, paragraph 8.

[7]   See table 8.3 for details <www.bis.org>.

[8]   See table 8.3 for details <www.fsb.org>.

[9]   See table 8.3 for details <www.iosco.org>.

[10]   See table 8.3 for details <www.fatf-gafi.org>.

[11]   See table 8.3 for details <www.iosco.org>.

(ii) Self-regulatory market operation

This second pillar comprises the networked market, constituted by the market participants, that is, the ETS administrators of participating jurisdictions and the entities authorized by them to trade inter-jurisdictionally, as well as the network entity operating and managing the transaction platform and its standing management body. As noted earlier,[12] there are models for how the network entity might be constituted – for instance, as a corporate entity with articles including the rules for participation and for market operation, to which jurisdictions joining would agree to adhere.[13] The participating jurisdictions would each hold a share in it. Alternatively, it may be an unincorporated association, with a charter to which participating jurisdictions subscribe, and with articles of association and procedural rules with which they agree to abide.[14] A question of the laws of which jurisdiction the network entity is created under will arise irrespective of what form it takes.

The standing management body of the network entity (for instance, its board of directors) would be responsible for governance and operational management direction.[15] It would provide information on the operation of the market to, and be subject to the supervision of, the supervisory bodies acting conjointly. In this sense, the market would not be entirely self-regulatory. In fulfilment of their roles and based on the analysis of the market and other data provided to them, the supervisory bodies might also provide guidance to the network entity to ensure efficient operation of the market does not impact global financial stability in any way and is effective in promoting and enhancing global mitigation efforts to achieve higher ambition.

Transactions carried out over the platform would be based on a set of standard terms and conditions. The network entity would be responsible for keeping the standard terms and conditions under review, to ensure their suitability and applicability to transactions in general. Specific transactions would proceed on the basis of a term sheet in which the variables applicable to that transaction, for example, counterparties, units being traded, price, and so on, would be set out.[16] Completion and verification of the information in the term sheet would effectively operate as conditions precedent to the transaction proceeding.

---

[12] In Chapter 7.

[13] Examples of what these rules might cover are set out, for illustration, in the Appendix.

[14] The FSB is a model along these lines with a charter, articles of association and procedural rules.

[15] Functions and structure of the standing management body were canvassed in Chapter 7.

[16] Illustrations of what variable elements might be included in a term sheet are set out in the Appendix.

Thus, once they were complete and verified, the transaction would automatically proceed to settlement and completion.

The market would be self-regulatory in the sense that participating jurisdictions would be responsible for its operation through the network entity and for funding that operation. Further, by inclusion of transaction rules into the digital code by which transactions are processed on the platform, any transaction proposed that did not comply, for example, because it would result in a net increase in permitted emissions, would not be able to proceed. Similarly, for any other applicable rules and conditions built into the code, whether rules of the network platform or conditions imposed by a jurisdiction on entities authorized by it to trade (for instance, a domestically imposed rule as to acceptable counterparties, such as only those from certain jurisdictions), non-observance would automatically mean no transaction.

(iii) Independent market information

The third pillar comprises the process and entities through which the outcomes of mitigation actions undertaken in different participating jurisdictions are valued.[17] The products of this process would be mitigation values that would attach to the units traded in the networked market. Thus, the function is to provide price sensitive information to the market and, as such, the sources of this information need to be independent, objective, credible, and reliable, and the process secure and trustworthy.

The concept of mitigation value and the mechanism by which assessment could be made has been canvassed earlier.[18] Further, it is noted that the idea of transparency and policy surveillance of countries in multilateral regimes is well established.[19] As noted earlier, Aldy and colleagues have reviewed a number of transparency models from other multilateral regimes. The conclusion is that, in terms of precedents for the proposed mitigation value assessment process, the

---

[17] The background to the process is set out in Chapter 7.

[18] In Chapter 7.

[19] Joseph Aldy, 'Designing a Bretton Woods Institution to Address Climate Change', 2012, HKS Faculty Research Working Paper Series RWP12-017, John F. Kennedy School of Government, Harvard University <https://dash.harvard.edu/bitstream/handle/1/8830777/RWP12-017_Aldy.pdf> accessed 17/04/18; Joseph E. Aldy, 'The Crucial Role of Policy Surveillance in International Climate Policy' (2014) 126(3–4) *Climatic Change* 279–92; Joseph E. Aldy, 'Evaluating Mitigation Effort: Tools and Institutions for Assessing Nationally Determined Contributions' Cambridge, Mass.: Harvard Project on Climate Agreements, November 2015 <http://pubdocs.worldbank.org/en/736371454449389076/pdf/Evaluating-Mitigation-Effort-Nov-2015.pdf> accessed 27/02/18.

international community can have resort to a range of transparency and policy surveillance models.[20]

Two possible approaches to how an assessment methodology might be applied are first, by a public, intergovernmental institution along the lines of the CDMEB model, emulating the models analyzed by Aldy.[21] Or second, by private sector entities under a regulatory model similar to that which is applied to credit reference agencies (CRAs) by the European Securities and Markets Authority (ESMA). As noted earlier, the latter approach is favoured.[22] In relation to how the regulatory body might be constituted, there are various models. The CDMEB again provides one such example, constituted under the Kyoto Protocol,[23] with details of its structure in decisions of the COP.[24] In terms of organizational structure, ESMA provides another model,[25] as a financial market regulator (particularly for regulation of CRAs).

2.   Seven Tiers of Governance

The governance structure for the market proposed can also be viewed horizontally as seven tiers, or layers of governance, illustrated in Table 8.2.

Considering these tiers from the bottom up:

(i) Conditions imposed by the jurisdiction electing to join

The bottom tier relates to the fact that transactions will be taking place between entities from individual, separate markets, each of which is continuing as an autonomous operation in its own right in its jurisdiction, while participating in the connection created by the networking arrangement. Thus, while an entity is trading only within its domestic market, the existence of the network is irrel-

---

[20] Ibid (Aldy 2015).

[21] Ibid (Aldy 2015) 18. Thus, also heeding the lessons Aldy draws out, such as the need to be a substantial, well-staffed, and well-functioning independent organization.

[22] A description of the suggested level of regulatory supervision for this approach is set out in the author's previous publication: Justin D. Macinante 'Operationalizing Cooperative Approaches Under the Paris Agreement by Valuing Mitigation Outcomes' [2018] *CCLR* 258. Reasons the author does not favour the CDMEB applying the assessment methodology are referenced there.

[23] Fn.4 (Article 12, paragraph 4).

[24] Fn.5 (Decision 17/CP.7).

[25] ESMA is established under the ESMA Regulation: Regulation (EU) No 1095/2010 of the European Parliament and of the Council of 24 November 2010 establishing a European Supervisory Authority (European Securities and Markets Authority), amending Decision No 716/2009/EC and repealing Commission Decision 2009/77/EC, OJ L 331, 15.12.2010, 84–119.

Table 8.2   Horizontal analysis: the seven tiers of governance

7. Overriding supervisory bodies, acting conjointly
6. Financial regulators
5. Market discipline
4. MV assessment process
3. Code for the transactions (smart contracts)
2. Network imposed conditions on jurisdictions joining
1. Jurisdiction electing to join network

evant to its trading activity and, in general, to the operation of that domestic market.

Once that entity seeks to trade outside its domestic market, with an authorized entity from another market, the rules and institutions of the networked market would become relevant and provide the legal framework within which that inter-jurisdictional transaction proceeds. In the first instance, those rules would include the conditions on which the jurisdiction authorizes an entity to engage in such inter-jurisdictional transactions, for example, by imposing conditions specifying a value range for acceptable MV for units that might be acquired from another jurisdiction. These conditions could be incorporated into the code for transactions entered by authorized entities from that jurisdiction so that, in the example, if the units proposed for purchase do not come within the range specified by that jurisdiction, the code would automatically prevent the transaction from proceeding.

(ii) Conditions imposed by the network on jurisdictions
The network entity will impose conditions on jurisdictions joining the network, relating to the operational and administrative requirements for participating in the network. For instance, these terms and conditions might include a commitment to funding a proportion of the network costs, or agreeing to abide by requirements as to permissioning for access to information on the ledger, or as to the consensus mechanism for adding verified information to the ledger. Other conditions might pertain to matters concerning eligibility requirements (for instance, as may be specified in the guidance on operationalization of Article 6), or to environmental integrity of transactions, or supplementarity requirements. Again, where applicable, these conditions could be incorporated into the code for transactions entered by entities authorized by the relevant jurisdiction so that, in instances where the conditions were not satisfied, the transaction could not proceed.

(iii) Code for transactions (smart contracts)

As envisaged herein, the contract between counterparties would be based on standard terms and conditions applicable to all inter-jurisdictional transactions on the distributed ledger platform, which in the case of an authorized entity from any particular jurisdiction would include the conditions imposed by the jurisdiction on entities authorized by it (as in (i) above) and transaction relevant conditions imposed on that jurisdiction by the network entity (as in (ii) above) that are applicable to entities authorized by it.[26]

The standard boilerplate terms and conditions would take the form of a master agreement similar in approach, for example, to that developed by the International Swaps and Derivatives Association (ISDA),[27] and by which the transaction counterparties would be bound. Counterparties would provide information pertaining to the particular transaction, such as details of the parties, units being transacted, price and other necessary variable details in the form of a term sheet, which together with the standard terms and conditions would constitute the contract between them. Once the term sheet information is complete and verified, the transaction would proceed to settlement and completion automatically and irreversibly.

(iv) The mitigation value assessment process

The mitigation value (MV) assessment process provides the value determined for the outcome of a mitigation action. Where the mitigation action is an ETS, this value is expressed as the mitigation value of a unit traded in that ETS. This value might be seen as the difference in mitigation with and without the action, adjusted for risk factors relating to the mitigation action itself, the suite of actions of which it forms part and the particular jurisdiction.[28]

By assessing mitigation actions to determine MVs for the outcomes, that is, the units traded in the ETSs of jurisdictions participating in the networked market, a common metric would be derived, enabling fungibility of the units across schemes. The MV provides a direct connection between the actual mitigation being achieved by these actions and market price of the outcomes. It also transmits information between counterparties about the respective jurisdictions. The MV of the units traded would be one of the variable elements included on the term sheet for a transaction.

---

[26] It is envisaged that these would probably be the same for all jurisdictions, for instance, representations and warranties that the jurisdiction satisfies eligibility requirements for engaging in cooperative approaches under Article 6 as agreed by the CMA.

[27] International Swaps & Derivatives Association, Inc., 2002 Master Agreement, as of June 9, 2010 <www.isda.org/about-isda/>.

[28] Fn.22 (Macinante).

(v) Market discipline on MV assessments

This level does not exist as any formal governance layer but rather in the broader sense of the governance structure. Nevertheless, market sentiment would operate as a reality check on the MV assessment process. The correlation between the mitigation value of a traded unit and its price is likely to lead to market price movements in any situation where the sentiment is that an MV assessment is not accurate. Thus, the market reaction on price will reflect a consensus on the MV assessed for any particular unit.

This might be viewed as a threat to the integrity of the market, as it could provide an opportunity for manipulation of the price. However, several considerations militate against such: first, if the networked market is successful engaging the private financial sector, it should be sufficiently deep and with a broad enough cross-section of participants as to make attempts to move the price for improper purposes unlikely to succeed; second, if the sources of MV assessment information are perceived to be independent, objective, credible, and reliable, not least because of the quality of the regulatory process under which that information is generated, then logically, an MV assessment would need to deviate significantly and obviously from the market expectation for traders to be willing to move against it; third, supervision of the domestic markets by financial regulators, supervision of the MV assessment process by the MV assessment regulatory body, both reporting to the overriding supervisory bodies, acting conjointly, should provide appropriately thorough and rigorous levels of scrutiny of all aspects of market activity as to make manipulation of this nature difficult to carry out successfully; and fourth, it is likely that market sentiment that an MV assessment was not accurate would manifest itself primarily in pricing of the futures contract for the carbon asset in question.

(vi) Regulators acting collaboratively

As noted, it is proposed that domestic financial regulators in each participating jurisdiction would monitor behaviour in the context of domestic market operation. In this respect, they would act collaboratively with their counterpart ETS administrator/regulator or, if Paris Agreement rules so provide, the jurisdiction's Designated National Authority (DNA). This domestic oversight would then feed into the oversight provided by the supervisory bodies.

(vii) The overriding supervisory bodies, acting conjointly

These bodies, one a climate subsidiary body established by the CMA under the Paris Agreement, the other an existing body, or a new committee or subordinate body of an existing inter-governmental financial organization, acting conjointly and reporting to the CMA, could be charged with setting overall policy direction, supervising the effectiveness of market operation in moving towards the climate objective, supervising the network of carbon markets behaviour as a global financial market, and supervising operation of the MV assessment regulator. They would advise, inform, and report on these matters to the CMA.

3.   Consideration of IPCC Criteria

Finally, as outlined in Chapter 5, the IPCC Fifth Assessment Report includes consideration of agreements and instruments for international cooperation in addressing climate change.[29] It proposes criteria to evaluate forms of international cooperation as: environmental effectiveness; aggregate economic performance; distributional and social impacts; and institutional feasibility.[30] These criteria are applied by the IPCC to different existing forms of international cooperation,[31] including the UNFCCC, Kyoto Protocol, the CDM, agreements under the UNFCCC pertaining to the post-2012 period, and other forms of international cooperation outside the UNFCCC, so there are parallels to the proposal set out here. For instance, elements of the governance structure for the market proposed here include, first, that it fosters pursuit of the climate policy objective by allowing for higher ambition, ensuring environmental integrity and transparency and applies robust accounting (environmental effectiveness); second, that it allows proper and efficient operation of the market through appropriate elements of financial regulation (economic performance), while the networked market should have similar cost-benefits to those ascribed to linking (cost effectiveness); and third, the functional separation of the self-regulating networked market would afford jurisdictions both a level playing field and relative ease in joining or leaving, based on their own domestic requirements (institutional feasibility). Thus, it could be claimed that the proposal compares favourably as a form of international cooperation, when considered in the same terms.

COMPARISON WITH EXISTING STRUCTURE

There is an obvious parallel between what is proposed here and the existing governance structure in that the Conference of Parties (COP) remains the

---

[29] Intergovernmental Panel on Climate Change (IPCC), 'International Cooperation: Agreements and Instruments' in Climate Change 2014: Mitigation of Climate Change. Contribution of Working Group III to the Fifth Assessment Report of the Intergovernmental Panel on Climate Change, [Edenhofer, O., et al. (eds.)]. Cambridge University Press <https://www.ipcc.ch/pdf/assessment-report/ar5/wg3/ipcc_wg3_ar5_chapter13.pdf> accessed 31/07/17.

[30] Ibid, 13.2.2.

[31] See, for instance, ibid, Table 13.3, 1042.

supreme policymaking body, decision-making body, and negotiating forum of the UNFCCC for both – for existing arrangements, as the Meeting of Parties to the Kyoto Protocol (CMP) and, with effect from December 2018, as the Meeting of Parties to the Paris Agreement (CMA). Yet, at the same time, these supplementary instruments to the Convention – the Protocol and the Agreement – mark the point of departure between the existing governance structure for emissions trading and that proposed.

1.   Differing Expressions as to Emissions Trading

Probably the most basic difference between the Kyoto Protocol (KP) and the Paris Agreement (PA) is the terminology used to refer to emissions trading. Under Article 17 KP, the COP has the role of defining the relevant principles, modalities, rules, and guidelines in particular, for verification, reporting, and accounting for emissions trading, which it has done in a series of decisions, starting with COP7.[32] On the other hand, Article 6 PA sets out requirements for the cooperative approaches and particularly for those involving the use of internationally transferred mitigation outcomes (ITMOs) towards nationally determined contributions (NDCs),[33] but requires them only to be consistent with guidance provided by the Subsidiary Body for Scientific and Technological Advice (SBSTA), pursuant to paragraph 36 of Decision 1/CP.21 as adopted by the CMA. The difference in terminology is clear – definition of principles, modalities, rules, and guidelines, as opposed to consistent with guidance – indicating a conceivably less prescriptive approach under the Paris Agreement. Yet questions remain whether the guidance (being developed under the Work Programme under the Paris Agreement (PAWP),[34] referred to as the Paris

---

[32]   Fn.5 (COP7). See for instance, Decision 19/CP.7 Modalities, rules and guidelines for emissions trading under Article 17 of the Kyoto Protocol. Also later decisions including, for example: 24/CP.8 (technical standards for data exchange), 11/CMP.1 (modalities, guidelines, rules for emissions trading), 13/CMP.1 (modalities, guidelines, rules for assigned amounts under Art.7.4 KP), 14/CMP.1 (standard electronic format for reporting), 16/CP.10 (issues related to registry systems under Art.7.4 KP).

[33]   Parties 'shall' promote sustainable development, ensure environmental integrity and transparency, including in governance, apply robust accounting to ensure, inter alia, avoidance of double counting: Article 6, paragraph 2.

[34]   UNFCCC COP23: Report of the Conference of the Parties on its twenty-third session, held in Bonn from 6 to 18 November 2017, Addendum, Part two, FCCC/CP/2017/11/Add.1, Action taken by the Conference of the Parties at its twenty-third session, 8 February 2018, I. Completion of the work programme under the Paris Agreement and Annex I, <https://unfccc.int/sites/default/files/resource/docs/2017/cop23/eng/11a01.pdf> accessed 23/01/19. Negotiators at COP24 and COP25 failed to reach agreement on relevant aspects: considered further in the following chapter.

Rulebook), in fact, will be less prescriptive and binding on parties in practice than is the case at present.[35]

2. Fundamentally Different Approaches

Second, there are fundamental differences between the PA approach to emissions trading and that taken in the KP beyond just the way they are expressed, which mean that, inevitably, the governance structure under the proposal will be different from that which exists at present. The KP differentiates between developed and developing countries in applying to developed countries quantified emission limitation and reduction commitments (QELRCs). These translate into assigned amounts and assigned amount units (AAUs), which along with other 'Kyoto units' could be surrendered and cancelled against emissions over a commitment period. Parties with these commitments are required to maintain a commitment period reserve (CPR) and there are eligibility and reporting requirements in order for a party to engage in emissions trading, which is only available to parties with QELRCs.

In contrast, while the PA differentiates between parties in terms of their respective capacities, it does not in terms of ability to engage in cooperative approaches (mitigation outcome transfers), which is open to all, although it has been reported that differentiation continues to be a contentious subject in the context of burden sharing in emissions reductions, given countries' different historical contribution to the causes and capacities to respond.[36] There are no QELRCs in the PA, but all parties are expected to put forward an NDC indicating, inter alia, the target emission level they will aim to achieve,[37] and these are to be periodically revisited[38] and the ambition increased.[39] Leaving to one side the sustainable development mechanism in Article 6, paragraph 4, there are no flexible mechanisms with corresponding units specified in the

---

[35] On the recurring issues of bindingness, prescriptiveness and differentiation, see: Daniel Bodansky and Lavanya Rajamani, 'The Issues that Never Die' [2018] *CCLR* 184.

[36] For instance: International Institute for Sustainable Development, Earth Negotiations Bulletin, Vol.12, No.733, Summary of Bangkok Climate Change Conference: 4–9 September 2018, 12 September 2018, 14 <http://enb.iisd.org/download/pdf/enb12733e.pdf> accessed 30/10/18. The differentiation debate is crystallizing around the scope of NDCs and, in terms of this proposal, may be relevant to how MV assessments are devised and undertaken. It is noted that the increased ambition in Article 4, paragraph 3 is expressed to reflect common but differentiated responsibilities and respective capabilities, in the light of different national circumstances.

[37] Article 4, paragraph 2.

[38] Article 4, paragraph 9.

[39] Article 4, paragraph 3.

PA. Rather, parties shall pursue domestic mitigation measures with the aim of achieving the objectives of their NDCs,[40] reflecting acceptance of the diversity and heterogeneity of approaches that countries may take. Article 6, paragraph 2, refers only to 'internationally transferred mitigation outcomes', as opposed to any specific unit that might be traded. The eligibility requirements for inclusion as, and questions of whether a specific value will be prescribed for mitigation outcomes (similar to the approach under the KP), are in the hands of the Paris Rulebook negotiators. On the latter point, the networked market clearly diverges by proposing independent, objective assessment of mitigation values, as evidenced in the third pillar described earlier.

Until the Paris Rulebook is fully agreed, it is difficult to provide more detailed distinguishing points, for example, in relation to matters such as requirements affecting eligibility to engage in international transfers of mitigation outcomes that will count towards a party's NDC and how they compare to the eligibility requirements for IET under the KP.[41] The negotiating text does include potential requirements, not dissimilar to those that applied under Article 17 KP.[42] All the same, some differences are readily apparent, for example, the accounting of assigned amounts is separated into three distinct phases under the KP, being the eligibility phase, the annual (trading) phase; and the end of commitment period phase when compliance is assessed.[43] In the networked market proposed this phased approach will not apply, not least because there is no assigned amount and possibly no commitment (or similar such) period, but also because the ledger will be continuously updated and accessible to appropriately permissioned entities. Eligibility criteria will be factored into the code for smart contracts such that transactions proposed by ineligible entities or from ineligible jurisdictions will be unable to proceed.

3.   Emphasis on Proposed Market as a Financial Market

A third point of distinction is that the proposal places greater emphasis on the inter-jurisdictional carbon market as a financial market. In a sense, this

---

[40]   Fn.37 (Art.4.2).

[41]   Although, note participation requirements in the Draft Text, paragraphs 3, 4, 5: UNFCCC CMA.2: Draft Text on Matters relating to Article 6 of the Paris Agreement: Guidance on cooperative approaches referred to in Article 6, paragraph 2, of the Paris Agreement, Version 3 of 15 December 00:50 hrs, Proposal by the President <https://unfccc.int/resource/cop25/CMA2_11a_DT_Art.6.2.pdf > accessed 15/01/20.

[42]   Ibid.

[43]   UNFCCC Kyoto Protocol Reference Manual on Accounting of Emissions and Assigned Amount, February 2007, 31 <https://unfccc.int/resource/docs/publications/08_unfccc_kp_ref_manual.pdf> accessed 23/10/18.

approach mirrors the bottom up approach often mentioned in relation to the PA,[44] since financial regulation is, in the first instance, a matter for domestic lawmaking. As the proposal is based on a network of autonomous domestic carbon markets, it is illustrative to consider briefly the approach taken in one such market, the EUETS being the obvious choice, since it constitutes the bulk of the global carbon market trading at present.

Illustration of EUETS

The EUETS was introduced on 1 January 2005 and covers energy-intensive industrial sectors, the power sector and as from 2012 the aviation sector. As at April 2018, Directive 2003/87/EC establishing a scheme for greenhouse gas emission allowance trading within the Union and amending Council Directive 96/61/EC (EUETS Directive)[45] has been amended by ten instruments, providing for matters such as linking with project-based mechanisms under the Kyoto Protocol; for all allowances to be held in a Union Registry, rather than in national registries of member states; and for the inclusion of aviation, amongst other matters.[46]

In particular, Article 12 of the EUETS Directive was amended in 2009[47] to include provision requiring the Commission, by 31 December 2010, to examine whether the market for emissions allowances was sufficiently protected from insider dealing or market manipulation and, if appropriate, to

---

[44] 'The Paris Agreement can be described as a hybrid between a top-down, rules-based system and a bottom-up system of pledge and review. The NDCs "codify" the bottom-up approach that emerged from Copenhagen': International Institute for Sustainable Development, Earth Negotiations Bulletin, Vol.12, No.663, Summary of the Paris Climate Change Conference, 29 November–13 December 2015, 43 <http://enb.iisd.org/download/pdf/enb12663e.pdf> accessed 26/06/17.

[45] OJ L 275, 25.10.2003, 32.

[46] Directive 2004/101/EC of the European Parliament and of the Council of 27 October 2004, OJ L 338, 13.11.2004, 18; Directive 2008/101/EC of the European Parliament and of the Council of 19 November 2008, OJ L 8, 13.1.2009, 3; Regulation (EC) No.219/2009 of the European Parliament and of the Council of 11 March 2009, OJ L 87, 31.3.2009, 109; Directive 2009/29/EC of the European Parliament and of the Council of 23 April 2009, OJ L 140, 5.6.2009, 63; Decision No.1359/2013/EU of the European Parliament and of the Council of 17 December 2013, OJ L 343, 19.12.2013, 1; Regulation (EU) No.421/2014 of the European Parliament and of the Council of 16 April 2014, OJ L 129, 30.4.2014, 1; Decision (EU) 2015/1814 of the European Parliament and of the Council of 6 October 2015, OJ L 264, 9.10.2015, 1; Regulation (EU) 2017/2392 of the European Parliament and of the Council of 13 December 2017, OJ L 350, 29.12.2017, 7; Directive (EU) 2018/410 of the European Parliament and of the Council of 14 March 2018, OJ L 76, 19.3.2018, 3; and Treaty of Accession of Croatia (2012), OJ L112, 24.4.2012, 21.

[47] Directive 2009/29/EC of the European Parliament and of the Council of 23 April 2009, OJ L 140, 5.6.2009, 63.

bring forward proposals to ensure such protection.[48] By a Communication dated 21 December 2010, the Commission provided a first assessment of the then current levels of protection of the carbon market from such misconduct and similar problems.[49] It reported that 75–80 per cent of the total volume traded in the EUETS was traded as derivatives contracts. The Commission noted the importance of information transparency, and canvassed the types of market abuse and other issues to be addressed. It noted that the then existing framework included financial markets legislation, the Market Abuse Directive (MAD) and the Markets in Financial Instruments Directive (MiFID), applying to emission allowances derivatives. Both of these items of legislation were under review at that time and there were a number of new financial markets measures proposed.[50]

The Commission reports periodically to the European Parliament and the Council on the functioning of the European carbon market.[51] In its 2017 report,[52] it noted that under the new MiFID II legislative package,[53] emission allowances (defined to include Kyoto project-based credits that are accepted for compliance purposes in the EUETS, as well as EU allowances) are classified as financial instruments, meaning that rules formerly applicable only to allowance derivatives also applied to the spot segment of the secondary carbon market, putting emission allowances on an equal footing in terms of transparency, investor protection and integrity.[54] Moreover, by virtue of

---

[48]   Article 12(1a).

[49]   European Commission, Communication from the Commission to the European Parliament and the Council, Towards an enhanced market oversight framework for the EU Emissions Trading Scheme, 21.12.2010, COM(2010) 796 final.

[50]   Ibid at 8. Also noted that a key future segment in the primary market, auctions, would come in full under the market oversight regime set out in the Auctioning Regulation: Commission Regulation (EU) No 1031/2010 of 12 November 2010 on the timing, administration and other aspects of auctioning of greenhouse gas emission allowances pursuant to Directive 2003/87/EC of the European Parliament and of the Council establishing a scheme for greenhouse gas emission allowances trading within the Community, OJ L 302, 18.11.2010, 1.

[51]   In accordance with Articles 10(5) and 21(2) of the EUETS Directive.

[52]   European Commission, Report from the Commission to the European Parliament and the Council, Report on the functioning of the European carbon market, COM/2017/693 final <https://eur-lex.europa.eu/legal-content/EN/TXT/PDF/?uri= CELEX:52017DC0693&from=EN> accessed 26/06/18.

[53]   Directive 2014/65/EU of the European Parliament and of the Council of 15 May 2014 on markets in financial instruments and amending Directive 2002/92/EC and Directive 2011/61/EU, OJ L 173, 12.06.2014, 394–496, took effect 3 January 2018.

[54]   Fn.52 (EC) 29. Note also that the MiFID II definition of emission allowances may need to be amended to accommodate international transfers of mitigation outcomes under the Paris Agreement, if the EU chooses to engage in cooperative arrangements under Article 6.

cross-references to the definition of a financial instrument, other financial market legislation such as the Market Abuse Regulation,[55] and the Anti-Money Laundering Directive[56] applied. Thus, the EU emissions trading market has been brought under financial market supervision.

Defining an emission allowance to be a financial instrument is not without problems, quite apart from the imposition it represents on the autonomy of EU member states' legal systems. It has been pointed out, for instance, that spot emission allowances differ from financial instruments in a technical sense, as they do not confer a financial claim against the public issuer, do not represent either title to capital or title to debentures and do not constitute forward contracts; from an application perspective, their primary purpose is to address climate change objectives, not to serve as an investment product; and from a regulatory perspective, invoking the legal obligations imposed by MAD and MiFID is onerous for smaller industrial enterprises.[57] Furthermore, legal and fiscal treatment of emission allowances varies across EU member states, with national treatment of allowances ranging from financial instrument and intangible asset to property right and commodity. These aspects – the legal and fiscal treatment – are not addressed in the EUETS Directive.[58]

Nevertheless, in spite of these variations, at the EU level the carbon market is treated as a financial market for regulatory purposes, suggesting this approach is seen as more effective and efficient than prior arrangements.

---

[55] Regulation (EU) No 596/2014 of the European Parliament and of the Council of 16 April 2014 on market abuse (market abuse regulation) and repealing Directive 2003/6/EC of the European Parliament and of the Council and Commission Directives 2003/124/EC, 2003/125/EC and 2004/72/EC, OJ L 173, 12.06.2014, 1.

[56] Directive (EU) 2015/849 of the European Parliament and the Council of 20 May 2015 on the prevention of the use of the financial system for the purposes of money laundering or terrorist financing, amending Regulation (EU) No.648/2012 of the European Parliament and of the Council, and repealing Directive 2005/60 of the European Parliament and of the Council and Commission Directive 2006/70/EC, OJ L 141, 5.6.2015, 73.

[57] Krzysztof Gorzelak, 'The Legal Nature of Emission Allowances Following the Creation of a Union Registry and Adoption of MiFID II – Are They Transferable Securities Now?' (2014) 9(4) *Capital Markets Law Journal* 373, 377, citing submissions on the consultation on MiFID review. Noted also that, in relation to provisions applicable as a result of emission allowance definition as a financial instrument, the UK Financial Conduct Authority has acknowledged that 'it is not always clear how all this overlapping legislation fits together': Financial Conduct Authority UK, The Perimeter Guidance Manual, Chapter 13, Guidance on the scope of MiFID and CRD IV, 13.4 Financial Instruments, Release 28, June 2018, at PERG 13/22 <https://www.handbook.fca.org.uk/handbook/PERG/13/4.pdf> accessed 02/07/18.

[58] Fn.52 (EC) 30. See also: fn.57 (Gorzelak). The issue of the precise nature of what is being traded is explored in detail in the following chapter (Chapter 9), when addressing specific legal issues, and also earlier (Chapter 4).

This should facilitate better investor protection in relation to areas such as market manipulation and insider dealing, and better investor and market risk management by drawing on risk management principles already developed for financial markets.

Greater emphasis on the networked market as a financial market is reflected in the proposal, first, by the introduction of the two supervisory bodies, one established under the CMA and the other established under an intergovernmental financial body such as IOSCO (although both reporting to the CMA), and acting conjointly; second, in acknowledgement that domestic financial regulators could play a bigger role in management of the market (as ESMA will do in the EUETS); and third, in the two other functional pillars, one being the self-regulatory market, and the other being the independent source of market information. These elements distinguish the proposal from the governance structure existing under the KP. The fact that the proposed market has a trading platform which will be owned, operated, and self-regulated by the participating jurisdictions is another point of distinction. Under Article 17 KP, there is no distinct trading platform, just bilateral agreements between counterparties, which are opaque as to terms such as price. Under the proposal a distinct inter-jurisdictional marketplace would be established to facilitate better price disclosure and better curtail improper market behaviour.

Separating the function of independently supplying market information further delineates the nature of the proposed market. By placing this assessment process on an objective, independent, structured, replicable basis it is intended to remove, to the greatest extent possible, the political element (although it is recognized that this will be difficult to remove entirely). Nevertheless, the process design would be intended to achieve, so far as is possible, a scientific outcome objectively, on a level playing field, not an outcome determined by compromise or political agreement. The proposal aims also to separate the functional process of deriving and delivering market information from the structural and operational aspects of the marketplace, thereby facilitating separation and, consequently, clearer resolution of the issues relevant to each function.

4.   Accounting and Informational Differences

Fourth, the KP accounting system is centred on two parallel information streams – GHG inventories and assigned amount information,[59] which starts at the national level. Each developed country (Annex I Party) is required to establish and maintain a national system for the preparation of its national

---

[59]   Fn.43 (KP Reference Manual) 37.

GHG inventory. On the assigned amount side, each Annex I Party is required to establish a national registry for tracking its holdings of and transactions of Kyoto units.[60] GHG inventory data and assigned amount information are compiled in national reports and are subject to review and compliance procedures. These procedures verify the Party's level of emissions and assigned amount, and its eligibility to participate in the Kyoto mechanisms.[61] Each Party's emissions and assigned amount information are recorded as official only after the information has been reviewed and any questions of implementation have been resolved through the compliance procedures. The Secretariat must maintain a compilation and accounting database (CAD) as the official repository of information related to each Party's accounting of emissions and assigned amount.[62]

The approach proposed here is for the networked market as a self-contained unit, so that there would be an information stream pertaining to the holdings and transactions of the authorized entities participating in trading. Information on account balances of those entities, at any point in time, would be available to the national administrators of participating jurisdictions. At the same time, the distributed ledger could be interrogated by the network entity for purposes of its own reporting to the supervisory bodies.

A fundamental difference is that under the KP, there has been no organized inter-jurisdictional trading platform, per se, that might be identified as 'the market', whereas the establishment of such a trading platform is proposed here. Thus, the market might be seen as largely self-contained and functionally separate from, but capable of feeding the required information into, other functional requirements such as overall NDC emissions accounting and reporting. Funding, administration, and operational responsibility would reside with the participating jurisdictions, through the network entity, which they would own and manage. In this way, only the countries that see a benefit in authorizing entities to trade inter-jurisdictionally contribute to the funding and maintenance of the market infrastructure. In the case of the KP, funding, administration, and operational management of the ITL and the CAD is through the Secretariat, therefore funded by all Annex I Parties.

5.   Compliance

A final aspect relates to compliance, in relation to which two points arise. First, emissions trading markets are not natural markets; demand needs to be

---

[60] Article 5, paragraph 1 KP; Decision 13/CMP.1, annex, paragraph 17 et seq.
[61] Article 7, paragraphs 1, 2; Article 8 KP.
[62] Decision 13/CMP.1, annex, paragraph 50 et seq.

supported by compliance and consequently, the threat of enforcement. Under the KP, international emissions trading is primarily directed to the Annex I Parties, thus the trading rules are backed up by the compliance mechanism.[63] Application of compliance and enforcement procedures against a sovereign party are always fraught with difficulty, as the transgressing sovereign party will have the ability to withdraw from the agreement.[64] In contrast, the proposed market and its operation are separated from the commitments made by participant jurisdictions through their NDCs, which in any case are voluntary. The networked market proposed is primarily based on the continued, autonomous operation of the carbon markets in the participating jurisdictions, thus compliance and enforcement will be primarily a domestic jurisdictional matter and so both more likely and more effective, providing a firmer underpinning to demand in the overall network.[65]

Second, under both the KP and this proposal, a non-compliant transaction or one involving non-compliant counterparties would not proceed. However, under the KP, the process is for the ITL and any relevant supplementary transaction log (such as the EUTL under the EUETS) to perform electronic checks on each transaction.[66] Under the proposal, the transaction process is designed so that a transaction cannot proceed unless the jurisdictions and the entities authorized by them to participate in the transaction are in compliance and the transaction, similarly, would not cause non-compliance (for instance, impacting environmental integrity by resulting in increased allowable emissions). The requirements are built into the code for the transaction, which will not proceed unless there is conformity with the requirements and that has been verified. The difference is that under the KP, the transaction involves a sequence of messaging steps[67] after the transaction is proposed, whereas in the proposal the code for performing the transaction automatically prevents the transaction from proceeding and alerts the counterparties to the non-compliance, thereby

---

[63]   Article 18 KP; Decision 27/CMP.1.

[64]   Canada withdrew from the KP at a time when it was unlikely to be able to meet its compliance obligations.

[65]   Along similar lines, a global federalist, bottom-up approach was advocated as early as 2005: D. G. Victor, J. C. House, and S. Joy, 'A Madisonian Approach to Climate Policy' (2005) 309(5742) *Science* 1820.

[66]   Fn.43 (KP Reference Manual) 69; UNFCCC Secretariat, Data Exchange Standards for Registry Systems under the Kyoto Protocol, Technical Specifications (Version 1.1.11), 24 November 2013, sections 4.6.1–5 (technical checks), 4.6.6–7 (policy and transactions checks) <https://unfccc.int/files/kyoto_protocol/registry_systems/itl/application/pdf/data_exchange_standards_for_registry_systems_under_the_kyoto_protocol.pdf> accessed 14/11/18.

[67]   Ibid (KP Reference Manual) 68, figure VI-6 (Sequence of Registry transactions).

removing the third-party gatekeepers, involving less process steps and so increasing efficiency of the process.

GLOBAL FINANCIAL MARKET GOVERNANCE STRUCTURES

A key aspect of the approach for global financial market governance that emerged following the global financial crisis of 2008 was that, '[A]s a supplement to sound micro-prudential and market integrity regulation, national financial regulatory frameworks should be reinforced with a macro-prudential overlay that promotes a system-wide approach to financial regulation and oversight and to mitigate the build-up of excess risks in the system'.[68] Thus, it has been observed that supervision over commercial actors in financial markets should be based on a two-tier system with national supervisors continuing to exercise micro-prudential oversight and a level of macro-prudential oversight introduced for financial markets as a whole in order to provide early recognition of systemic risks, although this would be more through enhanced cooperation of national authorities, rather than creation of a new global body.[69]

A similar two-level approach is proposed in the governance structure for the networked market. As noted above, with greater emphasis placed on the networked market as a financial market, it is proposed that domestic financial regulators would play a more important role, in a similar micro-prudential sense, in management of their respective emissions trading markets. In the first instance, this would be through the exercise of greater oversight as derivatives markets develop domestically. Second, if as in the EU, jurisdictions define the domestic allowance traded in their ETS as a financial instrument, this could be expanded to bring their spot allowance trading market under financial regulation (assuming domestic financial regulations that are similar to those in the EU), further enhancing this micro-prudential level oversight. The proposal is also that domestic financial regulators in those jurisdictions participating in the networked market contribute the membership of the overriding financial supervisory body that would act conjointly with the climate supervisory body. Together, these supervisory bodies would provide oversight at the macro-prudential level.

---

[68] G20 Working Group 1, Enhancing Sound Regulation and Strengthening Transparency, Final Report, March 25 2009, Executive Summary and Recommendations, ii <http://www.astrid-online.it/static/upload/protected/G20_/G20_wg1_25_03_09.pdf> accessed 20/11/18.

[69] Erik Denters, 'Regulation and Supervision of The Global Financial System' (2009) 1(3) *Amsterdam Law Forum* 63, 76–7 <http://amsterdamlawforum.org/issue/view/13> accessed 19/11/18.

While this approach to governance is consistent with developments in global financial governance, the later has not been without challenges. For instance, it has been noted that one fundamental underlying weakness in the international financial regime that remains is that there are too many institutions and mechanisms, with sometimes overlapping mandates, but limited powers.[70] The proposal set out in this book does not exacerbate it by proposing addition of another institution. Rather, it proposes that the financial supervisory body be a committee or subordinate body of one of the existing bodies in the global financial governance framework, as listed in the following Table 8.3.

As flagged earlier, IOSCO is considered the most appropriate of these bodies that might form a new committee to function as the financial supervisory body part of the governance structure proposed for the networked market. This conclusion is supported both by the nature of the functions performed by IOSCO and its existing committees in relation to regulation of financial markets, and by its composition and the broad coverage of that membership. All the same, the other bodies carry out functions that have a bearing on applications of DLT that are relevant to their areas of focus. For example, BIS-CPMI has undertaken analysis of the application of DLT in payment, clearing, and settlement to provide an analytical framework for central banks and other authorities to review and analyze DLT arrangements.[71] The responses of these and other organizations to applications of DLT are interrogated in the following section for how they might inform application to networking of carbon markets.

ANALYSIS OF RESPONSES TO DLT AND ITS APPLICATIONS

The surge of development related to information and communication technology in the decades immediately before and since the millennium has been characterized as the fifth technological revolution, the Age of Information and Telecommunication.[72] Yet while it is appropriate to include innovations such as distributed ledgers as part of the revolutionary developments, it is difficult to agree with commentators who see these technological developments as

---

[70] Council on Foreign Relations, 'The Global Finance Regime' Report by International Institutions and Global Governance Program, 2012 <https://www.cfr.org/report/global-finance-regime> accessed 06/07/18.

[71] Bank for International Settlements, CPMI, 'Distributed ledger technology in payment, clearing and settlement, An analytical framework', 2017 <https://www.bis.org/cpmi/publ/d157.pdf> accessed 18/10/18.

[72] Carlota Perez, 'Technological Revolutions and Techno-Economic Paradigms' (2010) 34(1) *Cambridge Journal of Economics* 185, 196–7.

Table 8.3   *Global financial governance regime bodies*

Body	Function	Composition
Bank for International Settlements	The BIS mission is to serve central banks in their pursuit of monetary and financial stability, to foster international cooperation in those areas and to act as a bank for central banks.	It is owned by 60 central banks, representing countries from around the world that together account for about 95 per cent of world GDP.
BIS-Committee on Payments and Market Infrastructures	CPMI promotes the safety and efficiency of payment, clearing, settlement and related arrangements, thereby supporting financial stability and the wider economy; monitors and analyses developments in these arrangements, both within and across jurisdictions. It also serves as a forum for central bank cooperation in related oversight, policy and operational matters, including the provision of central bank services.	CPMI representatives are senior officials of member central banks.
Financial Stability Board	FSB was established by the group of 20 industrialized countries (G20) with a key role in promoting the reform of international financial regulation. FSB operates through a three-stage process for the identification of systemic risk in the financial sector, for framing the financial sector policy actions that can address these risks, and for overseeing implementation of those responses	FSB is a not-for-profit association under Swiss law and is hosted by the BIS under a five-year renewable service agreement. The organization structure of the FSB consists of the Plenary, Steering Committee, Standing Committees, Working Groups, Regional Consultative Groups, Chair and the Secretariat. The Plenary is the sole decision-making body of the FSB. It consists of representatives of all Members and is currently composed of 54 representatives from 25 jurisdictions, six representatives from four international financial institutions and nine representatives from six international standard-setting, regulatory, supervisory and central bank bodies.

Body	Function	Composition
International Organization of Securities Commissions	IOSCO develops, implements and promotes adherence to internationally recognized standards for securities regulation. It works intensively with the G20 and the Financial Stability Board (FSB) on the global regulatory reform agenda.	IOSCO membership regulates more than 95 per cent of the world's securities markets in more than 115 jurisdictions; securities regulators in emerging markets account for 75 per cent of its ordinary membership. IOSCO committees cover: Issuer Accounting, Auditing and Disclosure; Regulation of Secondary Markets; Regulation of Market Intermediaries; Enforcement and the Exchange of Information and the Multilateral Memorandum of Understanding Screening Group; Investment Management; Credit Rating Agencies; Commodities Derivatives Markets; and Retail Investors, as well as Growth and Emerging Markets
Financial Action Task Force	FATF is an independent inter-governmental body that develops and promotes policies to protect the global financial system against money laundering, terrorist financing and the financing of proliferation of weapons of mass destruction. The FATF Recommendations are recognized as the global anti-money laundering (AML) and counter-terrorist financing (CFT) standard.	FATF currently comprises 35 member jurisdictions and two regional organizations, representing most major financial centres in all parts of the globe.

Notes: information from related websites (see footnotes 7–11 supra).

presaging entirely new systems of law.[73] For instance, it has been argued that the real innovation due to digital technologies '… is that, in the digital world, technology itself can be regarded as a parallel form of regulation. Such regulation derives from the technical features of various online platforms, which ultimately determine what can or cannot be done'.[74]

---

[73] Aaron Wright and Primavera De Filippi, 'Decentralized Blockchain Technology and the Rise of Lex Cryptographia' Background Paper, (Mar 12, 2015) Internet Governance Forum, UN-Department of Economic and Social Affairs, Workshops Descriptions and Reports, IGF 2015 Workshop No.239 Bitcoin, Blockchain and Beyond: FLASH HELP!

[74] Ibid, 46.

This has been described as Lex Informatica, an alternative normative system consisting of a particular set of rules and customary norms arising from the limitations imposed by design of the infrastructure subtending the network; a toolkit for regulation of online transactions through establishment of technical norms, in addition to contractual rules – the authors of that theory posit that this has led to establishment of a separate body of law.[75] They speculate that progressive deployment of blockchain technology may lead to recognition of another body of law – Lex Cryptographia, characterized by a set of rules administered through self-executing smart contracts and decentralized (and potentially autonomous) organizations.[76] These ideas are premised on 'cyberspace' being a separate (parallel) world, or jurisdiction, in which a different set of rules, or regulations, applies, giving rise to the potential for competition (or rather conflict) between the 'laws' of cyberspace and those of the real world.[77]

This book does not subscribe to the idea of a separate legal system for cyberspace, but rather roots its analysis of regulatory responses to DLT and its applications firmly in the existing legal world, beginning with the initial question of whether regulation should address the technology itself, or its applications. To examine this, a theory proposed in the literature for DLT regulation is examined in the next sub-section, before other regulatory analytical techniques are considered, in terms of applicability to the proposed model, in the sub-section thereafter. Evolving jurisdictional approaches to regulating applications of the technology are canvassed in the third sub-section.

1.   Regulation of the Technology or its Application

It has been observed that the '… patchwork of regulations applied to businesses using decentralized ledger technology is compromised by its inability to adapt to the technology, its inefficient mechanisms for responding to market and governance failures, and its overwhelming tendency to quash innovation in the name of preventing crime and protecting consumers'.[78] Thus, despite predictions of DLT revolutionizing the way things are done, that author, Reyes, argues that as a result of criminal and other illicit uses, regulators have adopted an increasingly aggressive approach to applying and enforcing exist-

---

[75]   Ibid, 48. See also Lawrence Lessig, *Code Version 2.0* (2nd edn, Basic Books, 2006).

[76]   Ibid.

[77]   Fn.75 (Lessig).

[78]   Carla L. Reyes, 'Moving Beyond Bitcoin to an Endogenous Theory of Decentralized Ledger Technology Regulation: An Initial Proposal' (2016) 61 *Vill. L. Rev* 191, 233. This focuses on responses by US regulators, both financial and criminal, from 2009, with the advent of bitcoin, onwards.

ing regulations against a different, new, and emerging technology, resulting in barriers to entry and a climate of legal stigma.[79] The literature (according to Reyes) tends to skip the question of how to regulate DLT and moves straight to jurisprudential questions of how blockchain might disrupt or alter known legal structures. In so doing, Reyes argues, a significant gap is left and DLT will never revolutionize contracts and so on, if the regulatory environment remains so hostile.[80]

Reyes concludes that there has been a failure to regulate at the DLT level, rather than just at the payments application level; an overly reactive approach to bitcoin market failures, AML and curbing illicit use applications; and that this focus is grounded in current characteristics of bitcoin and virtual currencies, thereby tying the regulation to a point in time. The problem is to find a regulatory approach that will treat DLT holistically, without tying regulations to specific applications.

Reyes' solution is an endogenous model of regulation that simultaneously governs from within and without by building compliance into the protocol, building on the idea of code-as-law, not as others have proposed,[81] but primarily directed at the technology itself. Thus, the idea is to regulate the technology itself by writing the regulation into the code, by '… leveraging smart contracts and other features of decentralized ledger technologies …'[82] However, this begs a number of questions, not least being how to distinguish between the code being regulated from the code of the smart contracts and other features. Perhaps more fundamentally, an obvious question is why regulate DLT? Or, equally fundamentally, what is DLT, as opposed to its applications, and does it (as opposed to the applications) actually need (or readily avail itself of) regulation? In other words, what is it that is being regulated when one regulates DLT? How, then, to regulate it?

The problem is that the technology is just lines of computer code, so for legal or regulatory purposes it does not, of itself, have a distinct economic or social function capable of being subject to legal framing (such as through regulation), but only derives such in its specific applications. There are coding rules that apply to how the code is written and, presumably, if these are not applied and observed the code will either not work, or will malfunction, or produce an undesired outcome.

This book argues that there is no tangible manifestation of the technology capable of regulation other than in the form of the various different appli-

---

[79]   Ibid. This is not necessarily the case in other jurisdictions, as is explored later in this section.

[80]   Ibid, 214.

[81]   Fn.73 (Wright and De Filippi).

[82]   Fn.78 (Reyes) 229.

cations. In most cases, the technology is being applied to scenarios that are already subject to regulation in some form or other, whether that is for mitigating systemic risk, for protecting consumers, or for preventing illegal or illicit activities. In applications where, by virtue of the technology, the reason for the regulation doesn't arise, for instance, by using the technology consumer risk does not arise, then there would be no reason to bring that application under the regulation that otherwise applies to the activity. There may also be instances where the applications are not covered by existing laws but, on proper consideration, give rise to public policy reasons why they should.[83]

This points to the need for case-by-case consideration on the part of the regulators, not holistic regulation of the technology. Such an approach is recommended in order to ensure that regulators do not stifle innovation in the underpinning technologies.[84] It has been pointed out that this approach aligns with core values of internet design and for this reason, most discussion of global internet governance has centred on higher-level use cases and prominent actors, leaving technical decisions on protocol specification to specialized standards bodies.[85] It has been noted also that: 'Regulators should focus on specific use cases of blockchains rather than the technology itself. This position finds support in experience with other disruptive technologies, such as the Internet and digital platforms'.[86]

2.   Regulatory Analytical Techniques

Notwithstanding the above, the way in which Reyes arrives at the endogenous theory of regulation for DLT is interesting: she follows a functional approach explored in financial regulatory literature.[87] Financial regulation is often tethered to the financial architecture – the design and structure of firms, markets, and other institutions at the time it is promulgated, but the financial system is changing dynamically.[88] Ongoing monitoring and updating can address

---

[83]   See for instance: Dirk A. Zetzsche, Ross P. Buckley, and Douglas W. Arner, 'The Distributed Liability of Distributed Ledgers: Legal Risks of Blockchain' (2018) 2018(4) *University of Illinois Law Review* 1361, 1382–3.

[84]   Julie Maupin, 'Mapping the Global Legal Landscape of Blockchain and Other Distributed Ledger Technologies', Centre for International Governance Innovation, CIGI Papers No.149, October 2017. <https://www.cigionline.org/sites/default/files/documents/Paper%20no.149.pdf> accessed 24/01/18.

[85]   Ibid, 5.

[86]   Michèle Finck, 'Blockchains: Regulating the Unknown' (2018) 19(4) *German Law Journal* 665, 689 citing Julie Maupin (fn.84 supra).

[87]   Reyes cites Steven L. Schwarcz, 'Regulating Financial Change: A Functional Approach' (2016) 100 *Minnesota Law Review* 1441.

[88]   Ibid, 1442.

this but is costly and prone to political interference, suggesting that it may be more effective, or at least instructive, to focus on the system's underlying, less time-dependent economic functions.[89]

Translating this approach to the governance structure for the networked market proposed here, the functions of this governance system might be couched in terms of, from a climate perspective, driving higher mitigation ambition towards limiting GHG emissions at levels that will confine global average temperature increase below the 1.5°C target and doing so by, from a market perspective, providing a stable global carbon price (or price range). A governance structure that focuses on these underlying functions would clearly be directed at the objectives of climate change policy.

Another regulatory analysis argues that only through a polycentric collaborative effort between industry and other stakeholders with regulators can the complex regulatory challenges of blockchain be satisfactorily addressed.[90] It proposes a number of guiding principles to facilitate achievement of that objective, being that: regulatory stability is a means of innovation and growth; public policy considerations must be considered from the outset; the importance of regulatory conversations; technological innovation triggers legal innovation; regulators should encourage experimentation; the focus should be on use cases rather than the technology; regulators should resist the temptation of prematurely creating new institutions; and regulators should engage in a transnational conversation.[91]

While these principles are directed to blockchain as a technical innovation more generally, nevertheless they have resonance in relation to the specific application of DLT envisaged by this book. Two, in particular, warrant consideration. First, technological innovation necessitates legal innovation; and second, focus should be on use cases rather than the technology. In relation to the first, Finck states that experience shows that while code is a self-regulatory mechanism, it should not operate in isolation from regulatory framing: a process of polycentric co-regulation acknowledges the limits of traditional top-down approaches in the context of technological innovation, while ensuring that public policy objectives are achieved.[92] The idea of co-regulation (also described as 'regulated self-regulation') encompasses various approaches in which the regulatory regime involves a complex interaction of general legislation and a self-regulatory body.[93]

---

[89] Ibid, 1444.
[90] Fn.86 (Finck) 685–7.
[91] Ibid.
[92] Ibid, 686.
[93] Ibid, citing C. Marsden, *Internet Co-Regulation: European Law, Regulatory Governance and Legitimacy in Cyberspace* (Cambridge University Press, 2011) 46.

The author, Marsden, addresses the origins of internet co-regulation as arriving at a typology of co-regulation and self-regulation. The various definitions canvassed[94] in the process are perhaps best covered by that of the European Community (as it then was): 'Co-regulation means the mechanism whereby a Community legislative act entrusts the attainment of the objectives defined by the legislative authority to parties which are recognised in the field (such as economic operators, the social partners, non-governmental organisations, or associations)'.[95] This idea of co-regulation translates neatly to the specific application of DLT envisaged in the self-regulatory market component of the governance structure. As outlined earlier, the various tiers of governance (levels 1–5 in Table 8.2), incorporating the climate policy (and legal provisions, assuming conformity with the Paris Rulebook) provide the legislative framework within which the self-regulatory market is entrusted to attain the objectives defined by the legislative authority, the CMA.

Second, as addressed in the preceding sub-section, the focus should be on use cases rather than the technology. Like the concept of co-regulation, this is a lesson derived from earlier experiences with the advent of the Internet.[96] This is demonstrated in practical terms by the evolving approaches to regulating applications of the technology, considered now in the sub-section following.

3.   Developing Jurisdictional Approaches

Distributed ledger technology is evolving and the range of potential applications, especially in relation to the financial sector, is expanding rapidly. Consequently, the response of legislators and regulators is in a state of flux, constantly reviewing developments and, increasingly, responding to technological changes and new applications with changes in applicable laws and in their approaches to regulating the use cases.

The initial application of blockchain as an alternative payment system (as bitcoin) has evolved in the decade or so since it first appeared, into fund raising through the issue of tokens, initial coin offerings (ICOs – which nomenclature

---

Marsden notes that this is often identified with the rise of 'new governance' in the late 1990s in environmental and financial regulation, but can be traced back to the inception of the Information Society policy in the mid-1990s.

[94]   Fn.93 (Marsden) 54–6.

[95]   Inter-institutional agreement on better law-making (2003/C 321/01), The European Parliament, The Council of the European Union and the Commission of the European Communities, Official Journal C 321, 31/12/2003 P. 0001–0005, Article 18.

[96]   Fn.86 (Finck) 689; see also fn.84 (Maupin). Note also, in particular: Raphael Auer, BIS Working Papers No.811, 'Embedded Supervision: How to Build Regulation into Blockchain Finance', September 2019, <www.bis.org> accessed 23/04/20.

is even evolving: now also known as 'token generating events' (TGEs), suggestive of changing emphasis and purpose), and investment vehicles. While many of the early concerns pertaining to the advent and use of bitcoin as an alternative payment system, such as its use for illegal or illicit transactions, tax evasion, anonymity of participants, money laundering, and terrorism financing risks continue, to these have been added fraud, hacking, and other consumer protection risks, and market manipulation and other market abuse issues as first bitcoin, then other subsequently issued digital coins (or 'virtual currencies') have rapidly become objects of arbitrage trading and investment. At the same time, the numbers of exchanges, platforms for trading, and other service providers (for example, 'wallet providers') have multiplied.

In its October 2018 submission to the UK House of Commons Treasury Select Committee report on crypto-assets, the Financial Conduct Authority (FCA) noted that in spite of technical limitations meaning that crypto-assets have not been able to scale up to challenge existing payment infrastructure, '… the crypto-asset market is developing at pace with over 1500 different coins and tokens valued at around \$311b'.[97] Both the House of Commons Treasury Select Committee report on crypto-assets[98] and the Securities and Markets Stakeholder Group (SMSG) in advice provided to the European Securities and Markets Authority (ESMA)[99] use the term 'crypto-asset' to encompass crypto-currency, virtual currency, virtual asset, and digital token; and 'token' rather than coin or currency, although others (see below) also use 'virtual asset'.[100] The SMSG advice includes background research showing significant changes in country of issuance for crypto-assets between 2017 and 2018, from the USA (32 per cent), Switzerland (27 per cent) and Singapore (21 per cent) in 2017, to Cayman Islands (40 per cent) and the Virgin Islands (21 per cent) in

---

[97] Financial Conduct Authority, 'Evidence on digital currencies to Treasury Committee (DCG0028)', House of Commons, April 2018, paragraph 1 <http://data .parliament.uk/writtenevidence/committeeevidence.svc/evidencedocument/treasury -committee/digital-currencies/written/81677.pdf> accessed 25/09/18. Five months later, the Securities and Markets Stakeholder Group advice to ESMA, citing the same source as the FCA, identified 1930 cryptocurrencies: see fn.99 (ESMA/SMSG) following.

[98] Treasury Committee, *Crypto-assets, (twenty-second report)* (2017–19, HC 910) <https://publications.parliament.uk/pa/cm201719/cmselect/cmtreasy/910/910.pdf> accessed 23/09/18.

[99] European Securities and Markets Authority (ESMA), Securities and Markets Stakeholder Group, 'Advice to ESMA Own Initiative Report on Initial Coin Offerings and Crypto-Assets', 2018, ESMA22-106-1338.

[100] Ibid, paragraph 12: 'The term "token" is more neutral as it does not carry the implicit legitimacy of "currency"'. Presumably the same argument can be made for using asset, opposed to currency or coin.

2018; in 2017, 78 per cent of the listed coins/tokens with a market cap of $50m or over, were found to be scams; 15 per cent continued to get traded, about half (7 per cent) of which were successful.[101]

Thus, the initial alternative payment system has transmogrified into fund-raising vehicle and source of investment assets, giving rise to a plethora of additional concerns, which are developing and evolving continuously. For instance, in other developments the Hong Kong Securities and Futures Commission (SFC), noting with concern the growing investor interest in gaining exposure to virtual assets via funds and unlicensed trading platform operators in Hong Kong, issued guidance on the regulatory standards expected of virtual asset portfolio managers and fund distributors,[102] while in a broader context, the Financial Action Task Force (FATF), noting the urgent need for an effective global, risk-based response to the anti-money laundering/counter-terrorism financing (AML/CTF) risks associated with virtual asset financial activities, has urged all jurisdictions to take legal and practical steps, such as ensuring that virtual asset service providers are subject to AML/CTF regulations, to prevent the misuse of virtual assets.[103]

(i) Survey of jurisdictions' responses

In these changing and challenging circumstances, this subsection briefly surveys the regulatory approaches in place, or planned, in illustrative jurisdictions.[104] The aim is to identify and examine some approaches being taken by legislators and regulators. The purpose is also, so far as possible, to elicit the direction in which regulation might move in the coming period in order to draw a picture of how a networked carbon market between jurisdictions oper-

---

[101] Ibid, paragraphs 21, 22.

[102] Securities and Futures Commission, Hong Kong, Statement on regulatory framework for virtual asset portfolios managers, fund distributors and trading platform operators, 1 November 2018, <https://www.sfc.hk/web/EN/news-and-announcements/policy-statements-and-announcements/reg-framework-virtual-asset-portfolios-managers-fund-distributors-trading-platform-operators.html> accessed 31/12/18.

[103] Financial Action Task Force, 'Regulation of Virtual Assets', 2018 <http://www.fatf-gafi.org/publications/fatfrecommendations/documents/regulation-virtual-assets.html> accessed 07/11/18.

[104] This does not attempt to be comprehensive, however, it is noted that the Library of Congress, Law Library, produced two overviews of cryptocurrency regulation in June 2018, the first covering 130 jurisdictions: Library of Congress, Law Library, Regulation of Cryptocurrency Around the World, June 2018 <https://www.loc.gov/law/help/cryptocurrency/cryptocurrency-world-survey.pdf> accessed 03/01/19; and the second, a selection of 14 jurisdictions: Library of Congress, Law Library, Regulation of Cryptocurrency in Selected Jurisdictions, June 2018 <https://www.loc.gov/law/help/cryptocurrency/regulation-of-cryptocurrency.pdf> accessed 03/01/19.

ating on a distributed ledger architecture, as proposed here, and the governance structure envisaged, might fit into this evolving environment.

(a) Typology of strategies

A typology of regulatory strategies for distributed ledger applications has been produced by Finck,[105] as follows:[106]

i. the wait-and see approach: regulators gather information, which is assessed, often in consultation with stakeholders and taking account of developments in other jurisdictions. The disadvantage of this approach is that until the regulator is in a position to classify the activity, innovators are faced with legal uncertainty with respect to the likely application of existing legislation;

ii. issue narrowing or broadening guidance on how existing legal frameworks apply: the disadvantage of this approach is that often the guidance is non-binding, thus leaving legal certainty lacking;

iii. regulatory sandboxing: 'defined as a set of rules that allows innovators to test their product or business model in an environment that temporarily exempts them from following some or all legal requirements in place'.[107]

iv. issuing new legislation: this approach is fraught with risk, not least in relation to changing terminology, as noted earlier in relation to the BitLicence in New York.[108]

v. using the technology for their own purposes: Finck acknowledges that this is not a regulatory strategy so much as an educational response, citing several jurisdictions where government agencies are partnering with technology providers to develop applications based on the provision of government services (government data availability in Ukraine, land registry in Sweden; inter-bank payments in Singapore).[109]

In practice, responses observed seem often to be a mixture of strategies, at times emanating from different parts of the same government,[110] for instance, a wait-and-see approach taken in conjunction with guidance on the application

---

[105] Fn.86 (Finck). It is noted that this is dated August 2017, meaning about three years of developments have taken place since.

[106] Ibid, 675–82.

[107] Ibid, 677.

[108] Angela Walch, 'The Path of the Blockchain Lexicon (and the Law)' (2017) 36(2) *Review of Banking & Financial Law* 713, 728.

[109] Fn.86 (Finck) 681.

[110] For example, in the UK, the Financial Conduct Authority, the Bank of England (includes Prudential Regulatory Authority (PRA)), HM Treasury, the Office of the Chief Scientist, HM Revenue & Customs (HMRC), and the House of Commons

of existing laws and a sandboxing initiative. The UK is a case in point: the Financial Conduct Authority (FCA) has established a regulatory sandbox to allow firms that may require authorization the ability to test products and services in a controlled environment, with restricted authorization and waivers;[111] and is participating in the Global Financial Innovation Network (GFIN), together with 11 financial regulators and related organizations, to create a 'global sandbox'.[112] The FCA also has a project to help innovator businesses understand the regulatory framework and how it applies to them.[113] As part of this project, the FCA has entered cooperation agreements with a number of other regulators to facilitate entry of innovative businesses into each other's markets, including Australia, Singapore, Hong Kong, Canada, Japan, Korea, and China.[114]

At the same time, the actions to be taken forward by the FCA, HM Treasury, and Bank of England (BoE) as outcomes of the House of Commons Treasury Select Committee report on crypto-assets, apart from continuing to monitor market developments and regularly reviewing the UK's approach (that is, wait-and-see), are otherwise aimed at shoring up the existing regulatory framework (thus, some guidance, some new legislation). They include: consulting on guidance for crypto-asset activities currently within the regulatory perimeter (FCA); consulting on a potential prohibition of the sale to retail consumers of derivatives referencing certain types of crypto-assets (for example, exchange tokens), including contracts for difference (CfDs), options, futures and transferable securities (FCA); consulting on potential changes to the regulatory perimeter to bring in crypto-assets that have comparable features to specified investments, and exploring how exchange tokens might be regulated if necessary (HM Treasury); transposing the EU Fifth AML Directive and broadening the scope of AML/CTF regulation further (HM Treasury); continuing to assess the adequacy of the prudential regulatory framework, in conjunction with

---

Treasury Select Committee are all active with respect to fintech, ICOs, and DLT applications in their respective roles.

[111] Financial Conduct Authority, UK, 'Regulatory Sandbox' <https://www.fca.org.uk/print/firms/regulatory-sandbox> accessed 31/12/18. Singapore, Switzerland, Australia, and even some US states are amongst other jurisdictions to have established regulatory sandboxes.

[112] Financial Conduct Authority, UK, Global Financial Innovation Network (GFIN), 7 August 2018 <https://www.fca.org.uk/print/publications/consultation-papers/global-financial-innovation-network> accessed 01/01/18.

[113] Financial Conduct Authority, UK, Innovate and Innovation Hub <https://www.fca.org.uk/firms/innovate-innovation-hub> accessed 31/12/18.

[114] Ibid.

international counterparts (PRA); and issuing revised guidance on the tax treatment of crypto-assets (HMRC).[115]

(b) Support for DLT applications
Jurisdictions also express statements of support for applications of DLT, especially in the financial sphere. For example, the October 2018 resolution of the European Parliament noted the potentially beneficial applications of DLT and urged the European Commission (EC), European supervisory authorities, and other institutions to investigate and develop applications in various sectors, including monitoring developing trends and use cases in the financial sector.[116] Another instance is the coming together of seven southern European states to sign the 'Southern European Countries Ministerial Declaration on Distributed Ledger Technologies' in December 2018. Cyprus, France, Greece, Italy, Malta, Portugal, and Spain express a vision to make southern Europe a leader in emerging technologies, such as DLT, and commit to share best practices with each other, while calling on the EC to continue the work it is undertaking through the European Blockchain Partnership.[117] They declare that legislation should allow innovation and experimentation in order that the public and private sectors better understand DLT and its use cases, and be based on European fundamental principles and technological neutrality.[118]

(c) Technology neutrality and regulatory guidance
The concept of existing legislation being applied on a technology neutral basis is used by many jurisdictions describing their approach to DLT use cases. For instance, 'Swiss legislation on financial markets is principle-based; one such principle is technology neutrality'.[119] German law has been described as being

---

[115] Fn.98 (Treasury Committee) Table 5.A.
[116] European Parliament, resolution 'Distributed ledger technologies and blockchains: building trust with disintermediation (2017/2772(RSP))', 3 October 2018, P8_TA-PROV(2018)0373. <http://www.europarl.europa.eu/sides/getDoc.do?pubRef=-//EP//NONSGML+TA+P8-TA-2018-0373+0+DOC+PDF+V0//EN> accessed 01/01/19.
[117] Southern European Countries Ministerial Declaration on Distributed Ledger Technologies, 4 December 2018, Brussels, Belgium <https://www.sviluppoeconomico.gov.it/images/stories/documenti/Dichiarazione%20MED7%20versione%20in%20inglese.pdf> accessed 01/01/19.
[118] Ibid.
[119] Swiss Financial Market Supervisory Authority FINMA, Regulatory treatment of initial coin offerings, FINMA Guidance 04/2017, 29 September 2017 <https://www.finma.ch/en/~/media/finma/dokumente/dokumentencenter/myfinma/4dokumentation/finma-aufsichtsmitteilungen/20170929-finma-aufsichtsmitteilung-04-2017.pdf?la=en> accessed 01/01/19. Others to make such statements include the UK FCA and Australian Securities and Investments Commission (ASIC).

'... generally agnostic as to the use of technology',[120] thus, there is neither specific DLT legislation, nor are there any express restrictions, but rather general German law principles apply.[121] The German context also provides an illustration of how guidance emanating from regulatory authorities might not always be straightforward. The German Federal Financial Supervisory Authority (BaFin) qualified bitcoin as being a 'unit of account' and thus a financial instrument within the meaning of the German Banking Act, meaning that engaging in commercial activities involving bitcoin without authorization under that act constitutes a criminal offence.[122] However, in September 2018, an appeal court in Berlin ruled that bitcoin does not qualify as a financial instrument for the purposes of the German Banking Act, as it did not represent units of account given that it lacks a stable value and is not an accepted means of payment. Nevertheless, BaFin is treating the decision as being limited to the facts of the case and maintaining its interpretation.[123]

---

[120] Jones Day, Lawyers, Blockchain for Business White Paper, November 2018 <https://www.lexology.com> accessed 19/11/18.

[121] Ibid. As such, German Federal Financial Supervisory Authority (BaFin) has followed a strictly no sandboxing approach, although some accommodation did apply in general based on the size of the company (principle of proportionality). The 'Blockchain Strategy' published by the German Government in September 2019 apparently now will apply regulatory sandboxes to permit testing of new technologies <https://www.bundesfinanzministerium.de/Content/EN/Standardartikel/Topics/Financial_markets/Articles/2019-09-18-Blockchain.html> accessed 27/01/20.

[122] Jens Muenzer, 'Bitcoins: Supervisory Assessment and Risks to Users' (2014) *BaFin Journal*, Expert article, <https://www.bafin.de/SharedDocs/Veroeffentlichungen/EN/Fachartikel/2014/fa_bj_1401_bitcoins_en.html> accessed 02/01/19.

[123] KG Berlin, Sept. 25, 2018, Docket No. (4) 161 Ss 28/18 (35/18), Court Decisions of Berlin-Brandenburg website; Gesetz über das Kreditwesen [Kreditwesengesetz] [KWG] [Banking Act], Sept. 9, 1998, Bundesgesetzblatt [BGBl.] [Federal Law Gazette] I at 2776, as amended, German Laws Online website <http://www.gerichtsentscheidungen.berlin-brandenburg.de/jportal/portal/t/279b/bs/10/page/sammlung.psml?pid=Dokumentanzeige&showdoccase=1&js_peid=Trefferliste&documentnumber=1&numberofresults=1&fromdoctodoc=yes&doc.id=KORE223872018&doc.part=L&doc.price=0.0#focuspoint> accessed 04/04/19. Also see: Library of Congress, Law Library, 'Germany: Court Holds That Bitcoin Trading Does Not Require a Banking Licence', *Global Legal Monitor,* 19 October 2018 <http://www.loc.gov/law/foreign-news/article/germany-court-holds-that-bitcoin-trading-does-not-require-a-banking-license/> accessed 03/01/19. Also fn.120 (Jones Day). The effects of this decision are limited to Germany in that 'unit of account' as a sub-category of the definition of 'financial instrument' is particular to the German Banking Act and it is the interpretation of 'unit of account' on which the court and BaFin differed.

(d) Case-by-case determination

Increased emphasis on ICOs/TGEs means that much of the regulatory focus is on how to address these activities so that they are not just a way of avoiding proper controls on fund-raising, but at the same time the controls imposed do not stifle innovation. At one end of the spectrum is the Chinese response, where on 4 September 2017, the People's Bank of China (PBOC) and six other regulators declared ICOs illegal and called on existing issuers to refund monies raised.[124] It has been reported that the Chinese authorities are ramping up the clampdown.[125] At perhaps the other end of the spectrum, the position adopted by the UK FCA, amongst others, is that 'Whether a crypto-asset (including crypto-tokens issued as part of an Initial Coin Offering) itself is capable of falling within the (regulatory) perimeter will … be fact specific depending on the particular crypto-asset instrument in question'.[126] All the same, as noted above, the House of Commons Treasury Select Committee report on crypto-assets does recommend authorities investigate further regulatory controls.

(e) Token taxonomy

The Swiss Financial Market Supervisory Authority (FINMA) guidance also indicates a case-by-case approach to the application of financial market law and regulation to ICOs, flagging areas of current regulatory law where the underlying purpose and specific characteristics of ICOs may intersect as being in relation to combatting money laundering and terrorist financing (issuing payment instruments); banking law (accepting public deposits); securities trading provisions (dealing in tokens that are securities); and collective investment scheme legislation (assets collected for external management).[127] Developing this approach, FINMA issued guidelines[128] focusing on the economic function and underlying purpose of the tokens, and distinguishing three general categories: first, payment tokens (synonymous with crypto-currencies) intended for use as a means of payment for goods or services; second, utility tokens that provide access digitally to an application or service; and third, asset tokens, representing a debt or equity claim on the issuer and thus analogous

---

[124] Fn.104 (Library of Congress/Selected Jurisdictions) 31.

[125] Fn.98 (Treasury Committee) paragraph 153. It is ironic that PBOC is also investigating issuing its own fiat crypto-currency: see fn.104 (Library of Congress).

[126] Fn.97 (FCA).

[127] Fn.119 (FINMA).

[128] Swiss Financial Market Supervisory Authority FINMA, Guidelines for enquiries regarding the regulatory framework for initial coin offerings (ICOs), 16 February 2018 <https://www.finma.ch/en/news/2018/02/20180216-mm-ico-wegleitung/> accessed 02/01/19.

to equities, bonds, or derivatives. Generally, asset tokens will be treated by FINMA as securities; if a payment token acts only as a means of payment and a utility token solely confers access to an application or service, then FINMA does not treat them as securities. The guidelines also recognize that these classifications are not mutually exclusive, allowing for tokens to have hybrid functionality.

The House of Commons Treasury Select Committee report on crypto-assets applies a similar typology in discussing crypto-assets, being: an exchange token (common uses as a means of exchange; to facilitate regulated payment services); security token (common use as a capital raising tool); and utility token (common use as a capital raising tool) and in the case of all three types, common uses include for direct investment or indirect investment.[129] Guidance from the Australian Securities and Investments Commission (ASIC), similarly, focuses on the indicators of when an ICO or token might fall within one of the definitions of a financial product.[130]

The SMSG advice to ESMA[131] applies the FINMA taxonomy as payment, utility, asset, or hybrid tokens to assess whether they are covered or should be covered by existing EU financial regulation. It concludes that payment tokens are not currently covered by MiFID II, the Prospectus Regulation, or Market Abuse Regulation. If they are transferable, they can be investment objects, in which case consideration should be given to listing them as a financial instrument under MiFID II. Similarly, utility tokens are not covered by financial regulation. Just as for payment tokens, if they are transferable, they can be investment objects, in which case the SMSG advises consideration should be given to listing them as a financial instrument.

In relation to asset tokens, the SMSG advice is more complicated. In order to determine whether financial regulations applied, it would be necessary to determine whether they are a financial instrument (for the purposes of MiFID II and the Market Abuse Regulation) and a transferable security (for the Prospectus Regulation). This depends on whether the asset token gives right to a financial entitlement, or an entitlement in kind (in which case, whether that includes a decision power in the project), and in both cases whether the token is transferable. Tokens giving right to an entitlement in kind without a decision power, but being transferable, might also share characteristics with derivatives, in which case questions arise as to whether the underlying asset is a commodity and, if so, whether cash settled or physically settled. Clearly,

---

[129] Fn.98 (Treasury Committee) chart 2.B.
[130] Australian Securities & Investments Commission, Initial coin offerings and crypto-currency, Information Sheet INFO225 <https://asic.gov.au/regulatory-resources/digital-transformation/initial-coin-offerings-and-crypto-currency/> accessed 25/09/18.
[131] Fn.99 (ESMA/SMSG) paragraph 47 decision-tree.

careful consideration of the structure and functionality of tokens is necessary to determine whether, and if so how, existing EU financial regulation applies.

(f) Instances of specific legislation

The SMSG advice[132] provides a desktop survey of legislative developments or regulatory approaches taken by national securities supervisory authorities in the EU, EEA Member States and Gibraltar, Switzerland, Channel Islands and Isle of Man, in regard of ICOs and crypto-assets, undertaken in August 2018. Seven jurisdictions (Malta, Switzerland, Lithuania, Gibraltar, Jersey, Isle of Man, plus France had proposals)[133] had expressly legislated or specifically developed methodologies, criteria or guidelines for assessing how and to what extent ICOs could be considered as financial instruments. Fifteen appeared to be taking the wait-and-see approach, dealing with proposals on a case-by-case basis, although as noted earlier, this can include a range of responses (Austria, Belgium, Bulgaria, Denmark, Estonia, Finland, Germany, Ireland, Luxembourg, Netherlands, Portugal, Spain, UK, Liechtenstein, Guernsey). The remaining 14 did not provide a clear position (Croatia, Czech Republic, Greece, Hungary, Italy, Latvia, Poland, Cyprus, Romania, Slovakia, Slovenia, Sweden, Norway, and Iceland).

While many governments are putting together taskforces and committees to examine the implications of new disruptive technologies, including DLT and its applications,[134] as the SMSG research bears out, most EU member states and other countries have not yet put in place legislation specifically relating to DLT, ICOs, or crypto-assets. Rather, amongst the jurisdictions reviewed, the SMSG report found there are 'very divergent regulatory approaches to crypto-assets'.[135] All the same, some jurisdictions already have specific legislation: for example, Japan amended its Payment Services Act in 2016 (effective 1 April 2017) defining 'cryptocurrency', requiring registration of and regulating cryptocurrency exchange businesses, and at the same time

---

[132] Ibid.

[133] French Parliament passed new ICO legislation in September 2018: White & Case LLP, 30 January 2019 <https://www.lexology.com/> accessed 05/02/19.

[134] For example, in December 2018, Israel announced the establishment of an interagency team for regulatory coordination in the area of virtual assets: Library of Congress, Law Library, 'Israel Establishes Interagency Team for Coordinating Regulation of Virtual Assets', *Global Legal Monitor,* 28 December 2018 <http://www .loc.gov/law/foreign-news/article/israel-establishes-interagency-team-for-coordinating -regulation-of-virtual-assets/> accessed 03/01/19.

[135] Fn.99 (ESMA/SMSG) paragraph 27.

requiring them to undertake AML checks. The provisions do not, however, cover ICOs.[136]

While the Japanese legislation was in response to cyber-attacks on unregulated exchanges, other jurisdictions' enactments are designed to attract tech business. For example, Gibraltar introduced the Distributed Ledger Technology Regulatory Framework on 1 January 2018, requiring locally-based firms using DLT on a commercial basis to store or transmit value belonging to others to be registered and adhere to a set of nine regulatory principles: it plans to bring ICOs within the ambit of the regulation.[137] In July 2018, Malta introduced three laws aimed at encouraging DLT projects to locate there: one of these, the Virtual Financial Assets Act 2018, regulates ICOs, requiring publication pre-issue of a white paper approved by an agent registered under the Act, who needs to be in place at all times to ensure compliance with the law.[138] Thirdly, Liechtenstein is moving forward with a Blockchain Act as part of its aim to take advantage of the potential of the technology, in an environment of legal certainty and user protection.[139] The Liechtenstein proposals allow for continuing technological evolution by creating a legal basis for much broader scope of application on a technology neutral basis (thus, 'trusted technologies' or TT systems, rather than just distributed ledgers or blockchain) in the 'token economy', where 'token' is a construct introduced to embody all types of rights on a TT system.[140]

Countries are also enacting provisions to tighten up the application of AML/CTF laws. The Law Library of Congress report of June 2018 identifies 17 countries that have applied AML/CTF legislation to DLT applications. For example, in Australia the Anti-Money Laundering and Counter-Terrorism Financing Act 2006 was amended in 2018 to require businesses providing convertible digital currency exchange services to be registered with and comply with mandatory reporting obligations to the Australian Transactions Reports

---

[136] Library of Congress, Law Library, 'Regulation of Cryptocurrency: Japan', June 2018 <https://www.loc.gov/law/help/cryptocurrency/japan.php> accessed 03/01/19.

[137] Fn.98 (Treasury Committee).

[138] Library of Congress, Law Library, 'Malta: Government Passes Three Laws to Encourage Blockchain Technology', *Global Legal Monitor*, 31 August 2018, <http://www.loc.gov/law/foreign-news/article/malta-government-passes-three-laws-to-encourage-blockchain-technology/> accessed 03/01/19.

[139] Principality of Liechtenstein, Ministry for General Government Affairs and Finance, Unofficial Translation of the Government Consultation Report and the Draft-Law on Transaction Systems Based on Trustworthy Technologies (Blockchain Act), 28 August 2018, LNR 2018-879, 36 <http://www.regierung.li/media/attachments/VNB-Blockchain-Gesetz-en-full-clean.pdf?t=636799366866600241> accessed 03/01/19.

[140] Ibid, 40.

and Analysis Centre (AUSTRAC).[141] European jurisdictions also have been legislating to transpose the provisions of the EU's Fifth Money Laundering Directive[142] into domestic law. For example, in the United Kingdom, the Money Laundering and Terrorist Financing Regulations (Amendment) Regulations 2019 were published in December 2019, while in Germany draft legislation was published in July 2019. These focus particularly on bringing crypto-asset service providers within the scope of money laundering provisions.

Although to date these legislative enactments (apart from AML/CTF) seem to be more the exception, rather than the norm, changes are perceptible that seem to impute a trend towards more targeted legislation. The final text for French legislation to provide, inter alia, for a regulatory framework for entities offering services in relation to digital assets, was adopted 11 April 2019.[143] This sets out a simpler, but protective regime for both issuers of ICOs and investors with the French Financial Markets Authorities as the single point of contact for issuers, applies AML/CTF requirements, a new legal regime for digital asset service providers, as well as holding out the lure of reduced corporate tax on certain gains, from 2022.[144]

Switzerland has introduced a new type of fintech banking licence (under Article 1b Swiss Federal Banking Act) as the third element of a three-pillar fintech programme, the two prior elements being an extension of the maximum holding period for third-party funds in settlement accounts from seven to 60 days; and introduction of a regulatory 'sandbox' creating an unregulated regime for small innovative projects. These both came into force in August 2017.[145] The new banking licence will permit companies that are not banks (for example, crowd lending platforms, trading platforms, payment service providers) to accept public funds up to CHF100 million, provided they are

---

[141] AUSTRAC <http://www.austrac.gov.au/digital-currency-exchange-providers> accessed 03/01/19.

[142] Directive (EU) 2018/843 of the European Parliament and of the Council of 30 May 2018 amending Directive (EU) 2015/849 on the prevention of the use of the financial system for the purposes of money laundering or terrorist financing, and amending Directives 2009/138/EC and 2013/36/EU, PE/72/2017/REV/1, OJ L 156, 19.06.2018, 43–74.

[143] Plan d'Action pour la Croissance et la Transformation des Entreprises (PACTE – Action Plan for Buiness Growth and Transformation) <https://www.gouvernement.fr/en/pacte-the-action-plan-for-business-growth-and-transformation> accessed 27/01/20.

[144] France's New Framework for ICOs and Tokens: Simple, attractive and protective, 4/2019 <https://www.economie.gouv.fr/files/files/2019/ParisEUROPLACE_FrancesNewFrameworkapril_2019.pdf> accessed 27/01/20.

[145] Confederation of Switzerland, 'Federal Council puts new fintech rules into force', Federal Council release, 5 July 2017 <https://www.admin.ch/gov/en/start/documentation/media-releases.msg-id-67436.html> accessed 04/01/19.

not engaging in typical banking activities, in other words, the funds may not be reinvested and no interest is to be payable on them, however, they must be held separately from the company assets, or booked so they are capable of identification at any time.[146] As financial intermediaries, the new licence holders will be subject to Swiss AML laws but benefit from market recognition attaching to prudential FINMA supervision, without being as highly regulated as traditional banking business.[147]

A further Swiss development is the release by the Federal Council, in December 2018, of a report on the regulatory framework for blockchain and distributed ledger technology.[148] The Federal Council 'wants to create the best possible framework conditions so that Switzerland can establish itself and evolve as a leading, innovative and sustainable location for fintech …'[149] It found that there was no need for fundamental adjustments to the Swiss legal framework, but rather just specific amendments, such as in civil law, increasing the legal certainty for the transfer of rights by means of digital registers: this involves distinguishing two types of tokens – those that primarily represent a value in the blockchain context such as crypto-currencies, and those that represent a legal position, such as a claim, membership, or a right in rem; in financial market infrastructure law, devising a new and flexible authorization category for blockchain-based financial market infrastructures; and in AML law, more explicitly anchoring the current practice of making decentralized trading platforms subject to AML legislation.[150]

(ii) Conclusions on responses

This short survey of jurisdictions illustrates regulatory approaches to DLT and use cases in the financial sector. It does not attempt to canvass, for instance, measures by jurisdictions that have or are investigating the introduction of crypto-currency in their own right; or different approaches being taken to taxation in this field across jurisdictions. Rather, the historical evolution of the use cases and application of DLT from an alternative payment system to the present emphasis on ICO/TGEs and investment, points to a changing and

---

[146]   Baer & Karrer, 'Legal framework for new Swiss fintech licence finalised – entering into force January 2019', Briefing December 2018 <https://www.lexology.com/> accessed 18/12/18.

[147]   Ibid.

[148]   Confederation of Switzerland, 'Legal framework for distributed ledger technology and blockchain in Switzerland An overview with a focus on the financial sector', Federal Council report, Bern, 14 December 2018 <https://www.newsd.admin.ch/newsd/message/attachments/55153.pdf> accessed 02/01/19. Note that as an executive body of the Swiss government, the legal analysis is not legally binding.

[149]   Ibid.

[150]   Ibid.

increasing array of risks for both users and regulators to countenance, while remaining focused on the potential opportunities technological innovation can provide. As a consequence, regulatory approaches and responses cover the gamut from wait-and-see, through supportive measures and guidance vis-à-vis current laws, to sandboxing, innovator-friendly regulatory frameworks, technology neutral risk-based application of AML/CTF, consumer protection, and systemic risk management provisions.

Even within the limited selection of jurisdictions surveyed by the SMSG, there are very divergent approaches to crypto-assets, making it difficult to discern any stand out direction that regulation might take apart from, perhaps, increasing in amount. All the same, it is noted that soon, potentially not so much technology neutral as more technology positive adjustments (witness the French legislation, and Swiss proposals) may be introduced to existing financial regulatory frameworks to encourage applications while, at the same time, making it easier for those applications to come 'within the fold' in terms of the usual financial regulatory expectations.

In the course of this evolutionary process, the obvious starting point from a regulatory perspective has been for jurisdictions to consider whether the financial sector applications of DLT fall within the ambit of existing financial regulation. As the FINMA guidelines, the SMSG analysis and the House of Commons Treasury Select Committee report on crypto-assets, amongst others, demonstrate, such analysis devolves into a question of the nature of the token issued, its economic purpose and function, and the rights and entitlements, if any, attaching to it. At the same time, it is noted that, at least in the EU, an emissions allowance has been defined to be a financial instrument under MiFID II.[151] Consequently, financial regulatory provisions apply to the EU carbon market, meaning that trading on a DL platform with tokens representing the units of emission allowance[152] would, in any case, be subject to financial regulation – but by definition, rather than because of their economic purpose and function (which is GHG emission mitigation), and the rights and entitlements attached thereto. In other jurisdictions where emission allowances or other mitigation outcomes are not yet defined as financial instruments, it would be necessary to address the question of their economic purpose and function and the rights and entitlements attached thereto, specifically to

---

[151] Directive 2014/65/EU of the European Parliament and of the Council of 15 May 2014 on markets in financial instruments and amending Directive 2002/92/EC and Directive 2011/61/EU, OJ L 173, 12.06.2014, 394, Annex I, Section C (11). Emission allowances are defined as any units recognized for compliance under the EUETS.

[152] Recalling also that emission allowance is defined as units accepted for compliance purposes in the EUETS and thus includes certified emission reductions that are so accepted.

determine whether trading on a DL platform using tokens would invoke local financial regulations. These and other legal issues that may be relevant to the governance structure for the networked market proposed are considered in the chapter following.

9. Analysis of the networked market – legal issues

The preceding chapter dissects the governance structure of the proposed networked market vertically into three functional pillars and horizontally into seven tiers of governance. This structure is examined first, by comparison to the existing structure under the Kyoto Protocol; second, by assessing it for compatibility with global financial market governance; and then by considering regulatory developments in relation to application of the technology. These developments warrant a more granular analysis, in this chapter, of the governance structure.

This chapter begins by examining the element that intersects all three areas of law, namely the nature of what will be traded in this networked market. The survey of jurisdictions in the last sub-section of Chapter 8 establishes that the nature of tokens, issued on distributed ledger platforms used in financial applications, is at the intersection between legal issues relating to applications of DLT and financial market regulatory issues. If the units traded on the network of markets – that is, as tokens traded on the DLT platform – represent units of mitigation value (MV), then the application of DLT and financial market regulation also intersect with the operation of climate law. The 'distributed ledger token/financial instrument/MV unit' is germane to all three. This point of intersection is examined in the first section of this chapter.

The second section analyzes other issues pertaining to the DLT application, such as potential conflict with data privacy laws, location of transactions, and dispute resolution; and then the third section canvasses issues that stem from the potential application of domestic financial regulation, considering the situation in the European Union (EU) as a specific illustration. The analysis concludes with a review of the position in relation to international climate law issues, which might arise from the development of rules for operationalizing the Paris Agreement, in relation to emissions trading under Article 6.

THE INTERSECTION POINT OF APPLICABLE LAWS

The governance structure analyzed in this book relates not to the mitigation action or carbon market of any jurisdiction in particular, but to a network of the markets of jurisdictions that choose to participate. The networked market sits above and thus, is external to, those markets and their regulatory and institutional frameworks and, as flagged in Chapter 7, it will not be the mitigation outcomes (for example, emission allowances, or project generated credits) from mitigation actions in the particular jurisdictions that are traded on the networked market, but rather a common metric (a 'vehicle' or transaction unit) representing the value of those mitigation outcomes as determined through a credible, independent, impartial MV assessment process. The transaction units, it is posited, representing the MV embodied in the mitigation outcome from which they are derived, would facilitate the transactions on the networked market by performing functions similar to those which an international currency serves for financial transactions between jurisdictions, that is, as a medium of exchange; as a unit of account; and as a store of value[1] (although in the international transaction context, the two former functions would be more important).

The transaction unit would serve as the vehicle for carrying out indirect exchanges between different types of mitigation outcome; and as the unit of account, it would define the rate of conversion from one mitigation outcome to another. It would also function as a transmitter of information: informing counterparties as to value (along with the price) so that they would not need to undertake time-consuming and expensive research of their own; and more broadly, conveying information about the performance of the market in policy terms.[2]

Notwithstanding that the governance structure being analyzed relates to the network sitting above the individual participating jurisdictions' markets, in considering the transaction unit, its nature, and how it would function, it is necessary to take account of both the network level, where transactions involving the transaction unit take place (in sub-section 2) and how transaction units might be treated for the purpose of domestic regulatory frameworks (in sub-section 3), since this is where the market participants and assets are principally regulated. First, however, there is the preliminary consideration of why it is necessary to have a transaction unit.

---

[1]   George S. Tavlas, 'The International Use of Currencies: The U.S. Dollar and the Euro' (1998) 35(2) *Finance & Development*, International Monetary Fund <http://www.imf.org/external/pubs/ft/fandd/1998/tavlas.htm> accessed 19/02/16. These functions are just the same as the functions of money, that is, the local currency in a domestic context.

[2]   See: Justin Macinante, 'Networking Carbon Markets – Key Elements of the Process', 2016, World Bank Group Climate Change, 22 <http://pubdocs.worldbank.org/en/424831476453674939/1700504-Networking-Carbon-Markets-Web.pdf> accessed 01/03/18.

1. The Necessity For a Transaction Unit

Why is a transaction unit necessary? Why not simply use another mitigation outcome such as a European emission allowance (EUA) or a Certified Emission Reduction (CER) to fulfil this transaction vehicle role, in the same way as the United States dollar (US$) or the Euro (EUR€) might be used as a transaction currency?

The answer is that while this certainly might be possible, there are reasons why it is considered unlikely. First, it is unlikely that a mitigation outcome from one jurisdiction, for instance, an EUA, could achieve sufficiently wide acceptance by other jurisdictions to be viable as a transaction vehicle. This is notwithstanding the fact that, in terms of market share, the EUA in the carbon market is probably comparable to the US$ in world trade. The reason, it is suggested, stems from the nature of the function it performs. The purpose of a transaction unit is to convey MV between the counterparties to a transaction. Notwithstanding that currently no definition has been formally sanctioned, MV might be thought of in terms of the physical amount of reduced or avoided greenhouse gas (GHG) emission to, or sequestered from, the atmosphere that can be attributed to a particular tradable unit under, or derived from, a GHG mitigation scheme.[3] The assessment of MV can be seen as a process whereby the mitigation outcome being assessed is compared to a theoretically perfect 100 per cent outcome of a mitigation action (that is, 100 per cent of the mitigation intended, projected, or claimed to be achievable, being achieved), in other words, comparison against a standard. The expression of the MV of a jurisdiction's mitigation outcome would be as a ratio to, or fraction of, the notional standard (perfect) outcome.[4]

As the European Emissions Trading Scheme (EUETS) has shortcomings (noted earlier), as most other jurisdictions' schemes would have as well, it is likely that political objections would be raised by other jurisdictions that neither an EUA, nor any other jurisdiction's mitigation outcome, could validly provide a suitable standard against which to measure the MV of their own mitigation outcomes. Second, and perhaps more importantly, given the nature and purpose for which mitigation outcomes – particularly emission allowances – are created, including the fact that they (emission allowances) are intended to reduce in number over time in accordance with the mitigation trajectories of their respective jurisdictions' economies, it might appear inappropriate for them to be fulfilling the transaction vehicle role.

---

[3] Justin D. Macinante, 'Operationalizing Cooperative Approaches Under the Paris Agreement by Valuing Mitigation Outcomes' [2018] *CCLR* 258, 260.

[4] Hence, for example, it might be expressed as a number between 0.00 and 1.00.

2.   Transaction Unit Treatment at the Network Level

Acknowledging that any existing mitigation outcome such as an EUA will not be suitable as a transaction vehicle, and thus that a transaction unit is necessary, or at least desirable, for carrying out transactions on the networked market, the question becomes what is a transaction unit, in a legal sense? Does it have, or need to have, a separate, distinct legal existence, or need only a notional existence for the purpose of facilitating transactions? The answer, it is posited, is a function of the transaction process. Consideration of that process suggests that, as a vehicle to facilitate transactions, the transaction unit may, but would not necessarily need to, have a separate legal existence.

For example, the transaction process might involve a series of steps, such as first, the transferor's mitigation outcome converted to transaction unit by applying MV; second, the transaction whereby transferor transfers transaction unit to transferee and transferee transfers consideration (payment) to transferor; and third, the transferee converts the transaction unit to a transferee mitigation outcome by applying MV. If these steps (that is, steps one, two, and three) were to flow consecutively and automatically from start to finish, the transaction unit might only be a notional value in that process to effect the conversion from one mitigation outcome to the other. On the other hand, this would not be the case if the transaction process were not to proceed automatically from start to finish. In the case where, say, the transferor might convert mitigation outcomes to transaction units (step one) and hold those units, perhaps awaiting favourable moves in the market price, or the transferee having taken receipt of the transaction units on completion of the transaction (step one), hold those units also possibly awaiting a favourable move in the market price, then the nature of the transaction unit would need to be reconsidered.[5] The third possible role of the transaction unit, that is, as a store of value, would become more relevant. The particular counterparty would be holding a unit that clearly has a value – the transaction unit represents an amount of MV that was embodied in the mitigation outcome from which it was converted. Thus, its legal nature would no longer be merely notional.

If the transaction process for the networked market were to allow for these individual steps, so that an authorized entity might hold transaction units in its account, what would be the legal nature of the units so held? In a sense, the transaction unit is just the same asset as the mitigation outcome from which it

---

[5]   It is noted that the favourable moves in the market in so far as they would pertain to changes in the MV which might alter the conversion rate would be subject to any rules relating to environmental integrity, and so this hypothetical situation might never eventuate.

is derived, since it is just a revaluation of that mitigation outcome against the standard (hence, a standardized value mitigation outcome). However, if this approach to the legal nature of the transaction unit were followed through, then a situation might arise in which transaction units would be of differing legal natures depending on the jurisdiction from which they derived, just as for the mitigation outcomes themselves. For example, assuming that emission allowances come within the definition of a mitigation outcome, there have been a variety of approaches across jurisdictions to defining the legal nature of emissions allowances.[6] In some jurisdictions they have been defined as intangible property, while others specifically exclude there being a property right attached; some jurisdictions deem them to be intangible assets or financial instruments, while others treat them as tradable commodities. Hence, while different jurisdictions have determined the definitional approach that works in their legal context, at the network level that diversity could not practicably carry through to the transaction unit – there would need to be clarity as to its specific legal nature.

In these circumstances, where there arises a need to define the legal nature of a transaction unit held in an account on the network platform, a practical approach could involve either (i) for the applicable law of the network platform – based either on the jurisdiction of, or choice of law governing, the network entity (as agreed to by participating jurisdictions) to be the basis for determining the legal nature of the transaction unit; or (ii) for the constitutional documents of the network entity and/or the transactional rules of the network, which as noted earlier form part of the governance structure for the network, to define the legal nature of the transaction unit (again, as agreed to by the participating jurisdictions). Either approach would afford a degree of certainty to transactions, to the resolution of potential disputes and to the legal entitlements of the entities holding transaction units in an account on the network, where they are determined to be held beyond the jurisdiction of any domestic regulatory framework. The next sub-section now considers how they might be treated when they do come within a relevant domestic regulatory framework.

3.   Transaction Units in Domestic Regulatory Frameworks

Continuing with the assumption that there is a transaction unit for the purpose of transactions in the networked market, it is necessary also to consider how transaction units might be treated for the purposes of domestic regulatory frameworks of participating jurisdictions. The following Figure 9.1 shows

---

[6]   These differences were reviewed in Chapter 4 supra.

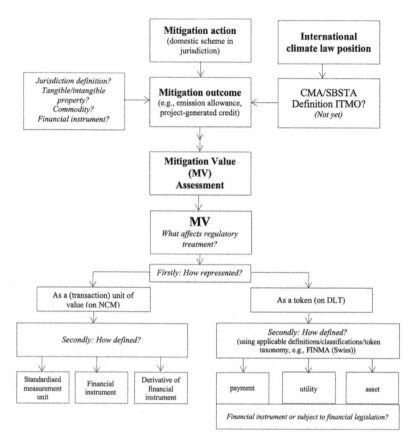

Figure 9.1   *Analysis-tree for domestic regulatory treatment of transaction unit*

an approach for making such a determination. It also illustrates how the three areas of law intersect in the mitigation outcome/financial instrument/token.

The progression illustrated is from the domestic mitigation action through mitigation value assessment to consideration of the outcome of that process in two ways: first, by asking how that assessed outcome, namely the MV, is represented – that is, as a legal instrument; and second, how that representation might be defined. In the networked market proposed, the possible instruments are either a transaction unit (under NCM), or a token (on DLT) or perhaps, both. The question is then whether the way in which they are defined invokes the financial regulatory regime in that jurisdiction and to what issues this gives rise.

To illustrate the application of this approach, the case of the EU is set out in Figure 9.2. In this case, the starting point is the EUETS, in which EUAs and certain CERs are accepted for compliance purposes. Under the updated Markets in Financial Instruments Directive (MiFID II),[7] emission allowances (defined to include Kyoto project-based credits that are accepted for compliance purposes in the EUETS (CERs), as well as EUAs) are defined as financial instruments.

There is no finally agreed definition of internationally transferred mitigation outcomes (ITMOs) agreed yet for the purposes of operationalizing Article 6 of the Paris Agreement,[8] so for these purposes it is assumed that EUAs and CERs, acceptable for compliance under the EUETS, will come within that definition. Hence, there are mitigation outcomes that are defined as financial instruments. If these mitigation outcomes are assessed to arrive at an MV, the question becomes what is the effect of applying the MV assessment to convert the mitigation outcomes into transaction units – does it make any difference to their treatment for regulatory purposes?

Even though the transaction units are essentially the same mitigation outcomes, but with a standardized value, such that it might be argued that they also should be considered to be a financial instrument, they are not. To be a financial instrument for the purposes of the EU financial regulatory framework,[9] they need to be listed in Annex 1, Section C of the Directive,[10] which in turn, could necessitate being acceptable for compliance purposes under the EUETS. They are neither listed in Annex 1, Section C, nor acceptable for compliance purposes (noting, of course, that presently they are only conceptual anyway).

There are two other potential ways in which the EU financial regulatory framework might be applicable: first, it is necessary to consider whether a transaction unit might be considered to be a derivative of the mitigation outcome (which is a financial instrument); and second, if the transaction units are represented by tokens on the EUETS platform, it would be necessary to

---

[7]   Directive 2014/65/EU of the European Parliament and of the Council of 15 May 2014 on markets in financial instruments and amending Directive 2002/92/EC and Directive 2011/61/EU, OJ L 173, 12.06.2014, 394, which took effect 3 January 2018.

[8]   See: UNFCCC: Draft Text on Matters relating to Article 6 of the Paris Agreement: Guidance on cooperative approaches referred to in Article 6, paragraph 2, of the Paris Agreement, Version 3 of 15 December 00:50 hrs, Proposal by the President <https://unfccc.int/resource/cop25/CMA2_11a_DT_Art.6.2.pdf > accessed15/01/20. Note that paragraph 1(e) currently provides ITMOs are generated in respect of or representing mitigation from 2021 onwards, which may, if agreed, alter this assumption.

[9]   The expression 'EU financial regulatory framework' is elaborated in the section following.

[10]   Fn.7 (MiFID II).

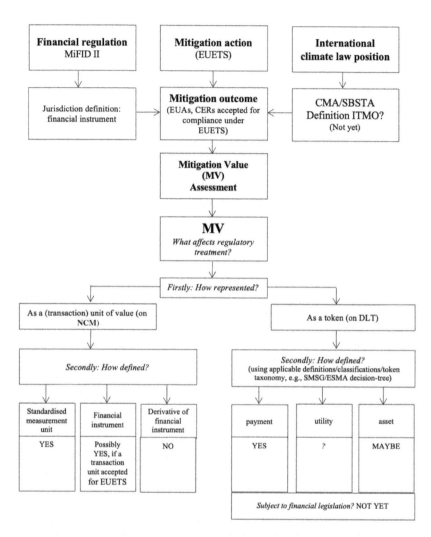

Figure 9.2   *Analysis-tree for domestic regulatory treatment of*
transaction unit applied to the case of the EU

assess whether the EU financial regulatory framework is invoked by virtue of the characterization of those tokens.

The first of these points can be dealt with shortly. A derivative is something that is derived from another source[11] and the transaction unit is derived from the mitigation outcome. However, in the sense considered here, namely as a financial derivative, a derivative is normally defined as a financial contract, the value of which is derived from the value of an underlying asset (the 'underlying').[12] The function of the derivative is to manage risks associated with movements in the price of the underlying[13] and the positions of the contract counterparties with respect these movements form the basis of the contract. Thus, while the value of the transaction unit can be said to be derived from the underlying mitigation outcome, clearly it is not a financial derivative but rather, as noted above, just a standardized value version of that mitigation outcome.

Second, if the transaction units are represented by tokens, it is necessary to consider the classification and identify characteristics of tokens that may invoke the EU financial regulatory framework. For this purpose, one might apply the approach elaborated by the Securities and Markets Stakeholder Group (SMSG) in advice provided to the European Securities and Markets Authority (ESMA).[14] The SMSG bases its taxonomy on the Swiss Financial Market Supervisory Authority (FINMA) guidelines[15] that distinguish three general categories: payment tokens intended for use as a means of payment for goods or services; utility tokens that provide access digitally to an application or service; and asset tokens, representing a debt or equity claim on the issuer and thus analogous to equities, bonds, or derivatives. In order to classify tokens on this basis and to determine the applicability of financial regulations, SMSG asked four questions: (1) does the token give the owner an entitlement against the issuer and, if so, what kind of entitlement? (2) Is it transferable? (3) Is it

---

[11]   *Concise Oxford English Dictionary* (11th edn (revised), Oxford University Press, 2006).

[12]   Michael Chiu, 'Derivatives Markets, Products and Participants: an Overview' in Bank for International Settlements (ed.), *Proceedings of the workshop "Data requirements for monitoring derivative transactions"* (Bank for International Settlements, 2012) <https://ideas.repec.org/h/bis/bisifc/35-01.html> accessed 15/01/19.

[13]   Ibid.

[14]   European Securities and Markets Authority (ESMA), Securities and Markets Stakeholder Group, 'Advice to ESMA Own Initiative Report on Initial Coin Offerings and Crypto-Assets', 2018, ESMA22-106-1338.

[15]   Swiss Financial Market Supervisory Authority FINMA, Guidelines for enquiries regarding the regulatory framework for initial coin offerings (ICOs), 16 February 2018 <https://www.finma.ch/en/news/2018/02/20180216-mm-ico-wegleitung/> accessed 02/01/19.

scarce and how is scarcity controlled? (4) Does it give decision power on the project of the issuer?

Applying the SMSG decision-tree questions to transaction unit tokens, the token does not give a right against the issuer;[16] it is transferable, but then the question is whether it has a use value: if it does, then the SMSG approach would treat it as a commodity, and if it doesn't, then the question would be is it scarce and how is that scarcity controlled? As the level of scarcity of transaction unit tokens would be changeable, the SMSG outcome would be to treat it as a currency (payment token). The SMSG conclusion is that neither commodities, nor payment tokens, are covered by the regulatory framework. However, as they are increasingly being held for investment purposes, raising concerns over investor protection and market abuse, SMSG suggests they should be defined under MiFID II as a financial instrument.[17] Thus, in the EU, while mitigation outcomes, that is, emission allowances (as so defined) are financial instruments under MiFID II, at present (were they to exist) neither a transaction unit nor a transaction unit token would be treated as a financial instrument, although the latter may be in future by reason of being considered a payment token.

In summary, this example illustrates two ways in which transaction units might potentially come within the ambit of domestic financial regulatory frameworks: either as a result of the mitigation outcome from which they derive being defined as a financial instrument (or otherwise subject to financial regulation) and the regulators recognizing that the transaction unit is essentially the same legal instrument but with a standardized value; or as a result of the domestic financial regulations applying to the transaction unit token by virtue of it being characterized as a payment token.

LEGAL ISSUES PERTAINING TO DLT

As the preceding chapter and the preceding section of this chapter underscore, analysis indicates that regulation of the distributed ledger technology on which it is proposed to build the network of carbon markets is very much a matter of the particular application proposed. All the same, there are issues of a legal nature that may be considered to relate more generally to use of the technology itself, not just specific applications. Mention of these issues was made when

---

[16] For the purposes here, it is not necessary to explore more technical design questions such as who would be that issuer, although for illustration, the network entity might be the issuer, issuing transaction units upon cancellation of the corresponding mitigation outcomes in the domestic registry by the transferring jurisdiction's registry administrator.

[17] Fn.14 (ESMA/SMSG).

introducing the technology in Chapter 6, some being inextricably linked to technical aspects such as system design, while others relate more to the structuring of legal relationships within that design. Nevertheless, it can be difficult to quarantine the legal aspects entirely from business and technical aspects, as technical issues concerning operation of the trading platform often will generate legal and governance issues. To illustrate this point, for example, network design is relevant to the speed of transaction processing and ledger updates for participants in different parts of the network, particularly geographically far-flung reaches of the network. At the same time, equivalence of access, first, to the market, and, second, to the same accurate, up-to-date market (ledger) information, is essential for ensuring a level playing field not just between trading entities, but also, critically, between participating jurisdictions (to generalize, for instance, lesser developed economies perhaps being more likely to be at the geographically farther flung reaches of the network). The design can influence time-lag issues, which in turn can influence participants' time critical access to market opportunities that could, conceivably, have legal ramifications.

As mentioned in relation to control over transactions,[18] contract execution on the DL network differs from existing third-party payment systems. However, the current state of development of DLT means that electronic processing speed is not as fast, nor is the transaction processing capacity anywhere near as great as, those third-party operated digital payment systems, such as MasterCard or Visa. This may be compensated, to some degree, by the end-to-end speed of the overall DLT-based transaction, since disintermediation of previously essential central counterparties and other intermediaries can reduce overall transaction time and cost. Whether and to what degree such disintermediation is realized in any particular transactions, however, will be affected by how the relevant jurisdictions' financial regulatory regimes apply.[19] Overall transaction speed is contingent also on the ability of the DL platform to provide an interface with a system for fiat currency settlement – possibly by a central bank or banks developing tokenized fiat currency. In the absence of such developments, alternatives would include: first, settlement using a crypto-currency (thereby introducing other additional value-related risks) or stablecoins;[20] second, by the platform operating on a 'centralized platform' basis, whereby matching and execution of orders, and corresponding transfer of assets and payment, all takes place on the platform but not on the

---

[18] Chapter 6, supra.
[19] This is considered in the next section.
[20] See discussion of stablecoins in Chapter 7, supra.

DL;[21] or otherwise, transactions needing to wait for payment processing to happen, effectively negating any potential time-cost benefit.

Other technical considerations that may give rise to legal issues (mentioned in Chapter 6) include: interoperability with existing systems and between systems; the absence of a recourse mechanism for dealing with mistakes; and problems due to erroneous coding, either in the way the smart contract code operates, or in the general functioning of the trading platform. To some extent, these considerations devolve to a question of whether the functions proposed are technically feasible or not. Assuming they are, then liability for errors of an operational nature with the coding must reside with the party whose responsibility it is (presumably on a commercial basis) to ensure that the coding is fit-for-purpose, irrespective of whether that question arises in the case of the code for a specific transaction or in the general operation and maintenance of the overall system. This will be a matter of the various contractual legal relationships between the technical contractors, the network entity running the platform, and the jurisdictional participants and their authorized entities using the platform. It is noted, in relation to contract coding that, as proposed, the contract terms would be standardized in the networked market, thus for individual counterparties it would only be a matter of entering the specific details of the transaction they wish to carry out. If they were to get that wrong, then it is considered that would be a failure of their own business management system, for which they should have to bear the liability risk.

As stated earlier, system design has a major impact on technical matters that can give rise to legal issues. For instance, design as to holding and updating the ledger by individual trading entities (nodes) is one technical feature with implications for the legal framework. The numbers of nodes holding and updating the ledger, or the extent of ledger information they hold, has implications for computing capacity and network scalability. This may be addressed technically (for instance, limiting to only the most recent ledger entries, or limiting the nodes, or limiting nodes to information only relevant to themselves), however, any such technical arrangement will need to fit within the parameters of the permissioning regime. The issue will arise also in respect of new entrants to the market. For instance, how much of the historic ledger they need to hold;

---

[21] European Securities and Markets Authority (ESMA), 'Advice on Initial Coin Offerings and Crypto-assets', 2019, ESMA50-1391, 44. This off-chain platform approach is seen by ESMA as risk prone due to the platform being a single point of attack for hacking, but conversely has the benefit of reducing congestion and scalability issues by settlement not being dependent on the DL. Counterparty risk in relation to the platform, also flagged by ESMA, would not be relevant to the networked market proposed here since the platform is proposed to be owned and operated by the participating jurisdictions. A version of this approach may prove effective in this context.

how they go about uploading it to their systems; how long it takes; and whether there are implications for their access relative to other earlier participants. These will all be relevant design considerations that, in turn, need to fit with the permissioning regime of who can view what information on the ledger.

Considerations of scalability and capacity of the ledger flag up another design issue, namely avoidance of choke points, where the network may be susceptible to interruption from internal technical problems or malicious external actions. This raises legal questions of liability for network performance and implications for parts of the network that may be isolated, even if only temporarily, the recourse that affected parties may have and who should bear responsibility. A pragmatic approach would be for jurisdictions to have the opportunity to raise concerns and consider their risk exposure and that of their authorized entities to such matters when deciding whether or not to join the network. Equally, the network entity operating the platform would then have the opportunity to take account of and act within its powers to address the concerns raised. Further liability issues arise in relation to trading entities whose actions, such as failing to safely maintain security of encryption keys, may precipitate interruptions. These issues, design affecting performance and participant actions that precipitate interruptions, could be provided for in the governance structure by appropriately framed rules: first, for jurisdictions that join the network; second, the conditions they impose on the entities they authorize to trade on the network; and third, the terms and conditions for transactions, built into the code for smart contracts.

Two final legal issues pertaining to DLT relate, first, to concerns over potential friction with participants' protection of confidential information and data privacy rights; and second, given the facility for cross-jurisdictional transactions on a DL network (and, in particular, the intended inter-jurisdictional trading proposed on the network of carbon markets), issues of jurisdiction, governing law of contracts, and appropriate forums for dispute resolution.

Concerns over the confidentiality of information held on the ledger, and how this might be balanced with, for instance, the Paris Agreement emphasis on transparency, relate to security and permissioning. As such, how this is addressed will be a function, again, of system design. The security aspect of design has been addressed earlier.[22] Permissioning rights, both for access to information concerning transactions or registry holdings, and in terms of interoperability with the ledger, are a function of design and will be guided by the legal requirements, such as those applying in relation to transparency and also in relation to reporting. Data privacy issues arise, inter alia, in relation to the permanence of the record on a DL system. One issue stems from the protec-

---

[22]   Chapter 6.

tions afforded personal data under data privacy protection laws, such as in the EU, the General Data Protection Regulation (GDPR),[23] the objective of which is to lay down rules, inter alia, relating to the protection of natural persons with regard to the processing of personal data.[24] The definition of 'personal data' in the GDPR, is:

> 'personal data' means any information relating to an identified or identifiable natural person ('data subject'); an identifiable natural person is one who can be identified, directly or indirectly, in particular by reference to an identifier such as a name, an identification number, location data, an online identifier or to one or more factors specific to the physical, physiological, genetic, mental, economic, cultural or social identity of that natural person; …

In theory there is the possibility of personal data being stored on the DL platform to the extent that a natural person is authorized to trade on the networked market, in which case the network entity would need to take account of the GDPR requirements, assuming the EUETS were to join. In practical terms, however, the likelihood of this causing a problem is small given that the DL network is unlikely to deal with natural persons, but only with transactions between business entities.

The permissioned nature of the network platform proposed, due to the fact that eligible participants will all need to have been authorized by their respective jurisdictions, means that questions of jurisdiction and governing law for resolving disputes should be addressed through design of the governance structure. To an extent, the potential for transactional legal disputes might be minimized by the fact that transactions cannot proceed unless all the preconditions have been satisfied and verified, at which point a contract should execute, settle, and complete automatically. Nevertheless, there will always be a possibility for legal disputes to arise, in which case the constitutional documents for the network entity, to which participating jurisdictions subscribe, might provide guidance either by specifying the applicable jurisdiction, law and forum for dispute settlement, such as recourse to some form of international alternative dispute resolution, or by providing a formula for determining such matters in any particular circumstances.

---

[23]   Regulation (EU) 2016/679 of the European Parliament and of the Council of 27 April 2016 on the protection of natural persons with regard to the processing of personal data and on the free movement of such data, and repealing Directive 95/46/EC (General Data Protection Regulation) OJ L 119, 4.5.2016, 1.

[24]   Article 1, paragraph 1, GDPR.

FINANCIAL MARKET REGULATION ISSUES

Determining when a transaction on the DL network platform might come within the ambit of the domestic financial regulatory framework of one or both jurisdictions of the counterparties to that transaction has been considered in the first section of this chapter. This section elaborates briefly what that might mean, by making a high-level survey of the current situation under EU law, by way of illustration.[25] In doing so, it is noted that the EU situation is complicated, in practice, by the differing national approaches transposing the provisions of EU law into national law. ESMA has expressed concern about the risks not covered by EU financial regulation posed by the growth in crypto-assets to investor protection and financial market integrity, most significantly through fraud, cyber attacks, money laundering, and market manipulation. But where they are covered, ESMA found a lack of consistency of interpretation, and lack of clarity in matters such as custody services and concepts of settlement and settlement finality across the EU.[26] At the same time, the European Banking Authority (EBA) has expressed the view that, even though crypto-assets and specific services such as custodian provision (that is, digital wallets) and trading platforms may typically fall outside the scope of EU financial regulation, divergent national approaches to regulating these activities are emerging, potentially giving rise not only to consumer protection, operational resilience, and market integrity issues, but also to level playing field issues.[27]

Both these European supervisory bodies have recommended further actions at the EU level, not only in relation to situations where the EU financial regulatory regime applies, but also for consumer protection where it does not, as well as broader application of AML/CTF requirements.[28] Nevertheless, the EU financial regulatory regime will apply when the token comes within the meaning of a financial instrument, as listed in Annex 1, Section C of MiFID

---

[25] Given that the networked market and transaction unit tokens proposed for trading in it are only conceptual, it is impractical to consider the application of the laws reviewed in a greater level of detail at this stage.

[26] Fn.21 (ESMA/ICO).

[27] European Banking Authority, 'Report with advice for the European Commission on crypto-assets', 2019 <https://eba.europa.eu/documents/10180/2545547/EBA+Rep ort+on+crypto+assets.pdf> accessed 11/01/19.

[28] Fn.21 (ESMA/ICO) Section VIII; fn.27 (EBA). Although, note implementation of the EU Fifth Money Laundering Directive: Directive (EU) 2018/843 of the European Parliament and of the Council of 30 May 2018 amending Directive (EU) 2015/849 on the prevention of the use of the financial system for the purposes of money laundering or terrorist financing, and amending Directives 2009/138/EC and 2013/36/EU, PE/72/2017/REV/1, OJ L 156, 19.06.2018, 43–74.

II.[29] This means '... a full set of EU financial rules, including the Prospectus Directive, the Transparency Directive, MiFID II, the Market Abuse Directive, the Short Selling Regulation, the Central Securities Depositories Regulation and the Settlement Finality Directive, are likely to apply to their issuer and/or firms providing investment services/activities to those instruments'.[30]

Notwithstanding that the token may come within the meaning of a financial instrument as defined in MiFID II, the application of each of the mentioned laws to the networked carbon market would specifically need to be considered. For instance, the Prospectus Directive[31] requires publication of a prospectus before a public offer of securities or admission of the securities to trading on a regulated market in the EU. The application of this Directive would depend on questions such as whether the tokens were securities (defined in Article 2(1) (a) as: '... transferable securities as defined by Article 1(4) of Directive 93/22/ EEC with the exception of money market instruments ...')[32] and whether the mechanism for the issue of tokens on the DL market platform could be considered to be 'an offer of securities to the public'. If the tokens only provide a mechanism for transactions between heterogeneous carbon markets, as considered above, it is unlikely they would be considered transferable securities and thus the Prospectus Directive would not apply. Even were they to become objects for investment by parties holding them to take advantage of market price movements, the fact that there is no financial entitlement such as a share of profit (hence equity) or pre-determined cash flow (as in a debt instrument) suggests they would not be subject to the Prospectus Directive.[33] Similar con-

---

[29]   Fn.7 (MiFID II).

[30]   Fn.21 (ESMA/ICO) paragraph 7.

[31]   Directive 2003/71/EC of the European Parliament and of the Council of 4 November 2003 on the prospectus to be published when securities are offered to the public or admitted to trading and amending Directive 2001/34/EC, OJ L 345, 31.12.2003, 64–89; amended by Directive 2010/73/EU of the European Parliament and of the Council of 24 November 2010 amending Directives 2003/71/EC on the prospectus to be published when securities are offered to the public or admitted to trading and 2004/109/EC on the harmonisation of transparency requirements in relation to information about issuers whose securities are admitted to trading on a regulated market, OJ L327, 11.12.2010, 1–12.

[32]   Transferable securities shall mean: shares in companies and other securities equivalent to shares in companies; bonds and other forms of securitized debt which are negotiable on the capital market; and any other securities normally dealt in giving the right to acquire any such transferable securities by subscription or exchange or giving rise to a cash settlement, excluding instruments of payment: Article 1(4) of Directive 93/22/EEC.

[33]   Fn.14 (ESMA/SMSG) paragraph 47.

siderations relate to the Transparency Directive,[34] which requires provision of information about issuers whose securities are traded on a regulated market in the EU. It would not apply unless the tokens traded on the networked market were classified as transferable securities.[35]

If the token were considered to be a financial instrument then MiFID II[36] (the Directive) and the Markets in Financial Instruments Regulation (the Regulation)[37] would apply. The Directive applies to investment firms, market operators, data reporting services providers, and third-country firms providing investment services or performing investment activities in the EU and establishes requirements for authorization, as to operating conditions, and in relation to the provision of investment services or activities.[38] As such, firms undertaking certain activities in relation to the tokens would need to be authorized. It is not necessary here to elaborate in detail the requirements under the Directive and Regulation, other than to observe that ESMA has expressed the view that where crypto-assets (and thus, potentially, tokens representing transaction units on the networked market) qualify as financial instruments, the trading platform may constitute a regulated market, or multilateral trading facility or organized trading facility thereunder.[39] Hence, jurisdictions intending to set up the networked market would need to weigh potential regulatory implications such as this when considering jurisdiction and governing law issues for the constitution of the network entity and the rules applicable in the networked market.[40]

The Market Abuse Regulation establishes a common regulatory framework on insider dealing, unlawful disclosure of inside information and market manipulation (market abuse) as well as measures to prevent market abuse

---

[34]   Directive 2013/50/EU of the European Parliament and of the Council of 22 October 2013 amending Directive 2004/109/EC of the European Parliament and of the Council on the harmonisation of transparency requirements in relation to information about issuers whose securities are admitted to trading on a regulated market, Directive 2003/71/EC of the European Parliament and of the Council on the prospectus to be published when securities are offered to the public or admitted to trading and Commission Directive 2007/14/EC laying down detailed rules for the implementation of certain provisions of Directive 2004/109/EC, OJ L 294, 6.11.2013, 13–27.

[35]   Fn.21 (ESMA/ICO) paragraph 101.

[36]   Fn.7 (MiFID II).

[37]   Regulation (EU) No 600/2014 of the European Parliament and of the Council of 15 May 2014 on markets in financial instruments and amending Regulation (EU) No 648/2012, OJ L 173, 12.06.2014, 84–148.

[38]   Article 1, MiFID II.

[39]   Fn.21 (ESMA/ICO) paragraph 106.

[40]   See preceding section, last paragraph.

to ensure the integrity of financial markets.[41] The Regulation would apply to tokens were they to qualify as financial instruments being traded on a regulated market, or multilateral trading facility or organized trading facility.[42] It is noted that as financial instruments, the Short Selling Regulation[43] would also apply to the tokens. However, it is envisaged that the mechanism for transactions on the networked market would not allow scope for short selling anyway, since it is proposed that the transferring counterparty would be required to verify holding the mitigation outcome units to be traded before the transaction could proceed.

The Settlement Finality Directive (SFD)[44] and the Central Securities Depositories Regulation (CSDR)[45] both have implications for settlement activities, if the DL platform were to be subject to their requirements. Most notably, this would be in terms of how settlement finality might be achieved on a DL platform, given the need for payment upon delivery and how a DL platform might interface with a system of fiat money for that purpose, also whether the DL platform could satisfy requirements for a central securities depository.[46] Another issue would be the extent to which central counterparties or clearinghouses might need to be interposed as part of the transaction process, by virtue of the SFD, countering the potential time and cost benefits that might otherwise be achieved by disintermediation of those roles as part of the transaction process. One approach to address these potential regulatory issues would be to build any prescribed regulatory roles into the trading platform itself, to create a hybrid structure.[47]

---

[41] Regulation (EU) No 596/2014 of the European Parliament and of the Council of 16 April 2014 on market abuse (market abuse regulation) and repealing Directive 2003/6/EC of the European Parliament and of the Council and Commission Directives 2003/124/EC, 2003/125/EC and 2004/72/EC, OJ L 173, 12.06.2014, 1–61, Article 1.

[42] Ibid, Article 2.

[43] Regulation (EU) No 236/2012 of the European Parliament and of the Council of 14 March 2012 on short selling and certain aspects of credit default swaps, OJ L 86, 24.03.2012, 1–24, Article 1.

[44] (Consolidated) Directive 98/26/EC of the European Parliament and of the Council of 19 May 1998 on settlement finality in payment and securities settlement systems, OJ L 166, 11.06.1998, 45–50.

[45] Regulation (EU) No 909/2014 of the European Parliament and of the Council of 23 July 2014 on improving securities settlement in the European Union and on central securities depositories and amending Directives 98/26/EC and 2014/65/EU and Regulation (EU) No 236/2012, OJ L 257, 28.08.2014, 1–72.

[46] The issue of the interface with the fiat money system discussed in preceding section.

[47] This issue has been discussed in Chapter 6; also: Stuart Davis and Julian Cunningham-Day, Linklaters LLP, 'Blockchain – recalibrating the market infrastruc-

Tokens may potentially fall also within the ambit of the Second Electronic Money Directive,[48] which defines electronic money to mean

> ... electronically, including magnetically, stored monetary value as represented by a claim on the issuer which is issued on receipt of funds for the purpose of making payment transactions as defined in point 5 of Article 4 of Directive 2007/64/EC, and which is accepted by a natural or legal person other than the electronic money issuer ...[49]

However, while the transaction unit tokens envisaged for trading on the networked market will satisfy most of the elements of this definition, the token as proposed would not represent a claim on the issuer and for this reason be unlikely to constitute electronic money.[50] Tokens may also come within the requirements of the Second Payment Services Directive[51] were they determined to be electronic money. However, as noted, this is unlikely in relation to transaction unit tokens as proposed. Finally, the fifth Anti-Money Laundering Directive[52] includes exchange services between virtual currencies and fiat currencies as well as custodian wallet providers as obliged persons for the purposes of anti-money laundering and counter-financing of terrorism requirements.[53] Given that both the European supervisory bodies mentioned, ESMA and the EBA, have recommended further actions at the EU level, these provisions are likely to be extended further, for example, to cover exchange services from one virtual currency to another.[54]

The foregoing brief review of the European domestic regulatory framework illustrates that characterization of tokens traded on the networked market as

---

ture', Going Digital Quarterly Breakfast Briefing, 14 October 2016, presentation Powerpoint slide deck.

[48]   Directive 2009/110/EC of the European Parliament and of the Council of 16 September 2009 on the taking up, pursuit and prudential supervision of the business of electronic money institutions amending Directives 2005/60/EC and 2006/48/EC and repealing Directive 2000/46/EC, OJ L 267, 10.10.2009, 7–17.

[49]   Ibid, Article 2(2).

[50]   See also first section of this chapter, penultimate paragraph.

[51]   Directive (EU) 2015/2366 of the European Parliament and of the Council of 25 November 2015 on payment services in the internal market, amending Directives 2002/65/EC, 2009/110/EC and 2013/36/EU and Regulation (EU) No 1093/2010, and repealing Directive 2007/64/EC, OJ L 337, 23.12.2015, 35–127.

[52]   Directive (EU) 2018/843 of the European Parliament and of the Council of 30 May 2018 amending Directive (EU) 2015/849 on the prevention of the use of the financial system for the purposes of money laundering or terrorist financing, and amending Directives 2009/138/EC and 2013/36/EU, PE/72/2017/REV/1, OJ L 156, 19.06.2018, 43–74.

[53]   Ibid, Article 1(1)(c).

[54]   Fn.21 (ESMA/ICO) Section VIII; fn.27 (EBA) chapter 3.

financial instruments may give rise to a number of domestic financial regulatory implications. These range from prospectus requirements and disclosure of
information about issuers, to operational rules for the providers of investment
services and activities, to rules prohibiting insider trading, or unlawfully
disclosing inside information, or manipulating markets; through to the end of
the transaction process, and rules relating to clearing and settlement finality.
Whether and to what extent similar such provisions might apply in the case
of any other domestic jurisdiction participating in the networked market, will
depend not only on how the transaction process on the DL trading platform
is designed and the function performed by the tokens, but also on how those
domestic regulatory provisions are framed.

CLIMATE CHANGE LEGAL ISSUES

Climate law issues that will arise for the governance structure for the networked market principally arise from the developing rules for operationalizing
cooperative approaches under Article 6, paragraph 2, Paris Agreement. These
were to be set out in the form of guidance to be considered and adopted by the
Conference of Parties serving as the Meeting of Parties to the Paris Agreement
at its first session (CMA.1); however, the Parties were unable to reach agreement at either first or second sessions, resulting in postponement for consideration and adoption to CMA.3.[55]

Notwithstanding that inconclusive and continuing process, potential issues
can be gleaned from the text before negotiators (the 'Draft Text'), which comprises a draft decision and Annex setting out the draft guidance.[56] The Annex
has seven chapters, which provide for internationally transferred mitigation
outcomes (ITMOs); rules for participation in cooperative approaches; the
application of corresponding adjustments; reporting; review; recording and
tracking; and for ambition in mitigation and adaptation actions. Governance is
not addressed directly as in previous iterations, rather the draft provides that
an Article 6 expert technical review team shall review the consistency of the
information submitted pursuant to the reporting obligations (Annex, chapter
IV) with the guidance,[57] in accordance with modalities, procedures and guidelines adopted by the CMA.[58]

---

[55]   UNFCCC CMA.2: Report of the Conference of the Parties serving as the meeting
of Parties to the Paris Agreement, second session, Madrid, 2–13 December 2019,
FCCC/PA/CMA/2019/L.9   https://unfccc.int/sites/default/files/resource/cma2019
_L09E.pdf accessed 15/01/20.

[56]   Fn.8 (UNFCCC Draft Text, Art. 6, para. 2).

[57]   Ibid, paragraph 25.

[58]   Ibid, paragraph 26.

The definition of ITMOs provides, inter alia, that they are: real, verified, and additional; emission reductions and removals, including mitigation co-benefits; measured in tonnes of carbon dioxide equivalent in accordance with IPCC methodologies and metrics adopted by the CMA (or other non-GHG metrics determined by the participating parties that are consistent with their NDCs); from cooperative approaches authorized pursuant to Article 6, paragraph 3; and generated in respect of or representing mitigation from 2021 onwards.[59]

Chapter II of the Annex (paragraphs 3, 4, and 5) sets out responsibilities for Parties participating in cooperative approaches, listing matters including being a party to the Paris Agreement;[60] having prepared, communicating, and maintaining a nationally determined contribution (NDC);[61] and having authorized and obtained authorization from other participating Parties to use internationally transferred mitigation outcomes (ITMOs), pursuant to Article 6, paragraph 3, Paris Agreement;[62] having arrangements for tracking ITMOs;[63] and having provided its most recent national inventory report.[64] In so far as they relate to the governance structure, these and other matters listed, or that may in future be listed under this heading, could be seen as constituting pre-conditions to be satisfied by jurisdictions wishing to join the network.[65]

More directly pertinent is Recording and Tracking (Annex, chapter VI), which provides for parties participating in cooperative approaches involving the use of ITMOs to have or have access to a registry for the purpose of tracking,[66] the secretariat being required to implement an international registry for those parties that do not have access to one.[67] This international registry is to be part of the centralized accounting and reporting platform (Annex, chapter VI.C) that the secretariat is required to implement 'to publish information submitted by participating Parties pursuant to chapter IV'.[68] Also as part of the centralized accounting and reporting platform, the secretariat shall implement an Article 6 database (Annex, chapter VI.B), to enable, inter alia, recording of corresponding adjustments and emission balances for and information on actions taken in relation to ITMOs by participating parties.[69] The secretariat is required also to maintain public information on ITMOs and publicly available

---

[59]   Ibid, paragraph 1.
[60]   Ibid, paragraph 4(a).
[61]   Ibid, paragraph 4(b).
[62]   Ibid, paragraph 4(c).
[63]   Ibid, paragraph 4(d).
[64]   Ibid, paragraph 4(e).
[65]   See, in this respect, Chapter 7 supra and the Appendix.
[66]   Fn.8 (UNFCCC Draft Text, Art. 6, para. 2), paragraph 29.
[67]   Ibid, paragraph 30.
[68]   Ibid, paragraph 35.
[69]   Ibid, paragraph 32.

information submitted by participating parties, and report annually to the CMA on these matters.[70]

These chapter VI elements of the Draft Text, Annex add clarity to the governance provisions evolving from the negotiations. The secretariat's role in maintaining a centralized accounting and reporting platform that includes both an international registry for parties that do not have a registry and an Article 6 database, of itself does not exclude the possibility of a networked approach to trading along the lines proposed in this book, which could well dovetail into such governance arrangements. Nevertheless, it might be observed that it does raise the spectre of carbon trading continuing in its current framing, compartmentalized in the climate regime. All the same, in making this observation it is noted that until the Paris Rulebook is finalized it is not helpful to delve into more detail: the Draft Text may not be the form of guidance to be considered and adopted by CMA.3 (or later), which ultimately, may very well be different.

Reports of the negotiations that led to the inconclusive outcome for CMA.1, point to an impasse over how to avoid double counting and to make adjustments for mitigation outcomes transferred internationally.[71] Reports of CMA.2 indicate differences over the application of corresponding adjustments; the future use of Kyoto Protocol units; application of a Share of Proceeds deduction to transactions for application to the Adaptation Fund (as under the CDM); and ensuring overall mitigation through cancellation of ITMOs.[72] A group of 31 countries, led by Costa Rica and Switzerland, launched a set of principles, the San Jose Principles for High Ambition and Integrity in International Carbon Markets, as constituting the basis on which a fair and robust carbon market should be built, amongst other things emphasizing environmental integrity to enable the highest possible mitigation ambition. It was reported also that the EU had stressed that failure to agree the rules for Article 6 did not prevent domestic or international carbon markets.[73]

Whether these events portend widening gaps between positions and more delay, or flag resolve and commitment to carbon markets is unclear. Yet it serves to suggest that the nature of governance arrangements – and, for that matter, most other issues that could militate against the networked market proposed here – are probably not so much technical, as political.

---

[70] Ibid, paragraph 36.

[71] International Institute for Sustainable Development, Earth Negotiations Bulletin Vol.12 No.747, Katowice Climate Change Conference, 18 December 2018, Summary and Analysis, 32–33 <http://enb.iisd.org/climate/cop24/enb/> accessed 18/12/18.

[72] International Institute for Sustainable Development, Earth Negotiations Bulletin Vol.12 No.775, Summary of Chile/Madrid Climate Change Conference: 2–15 December 2019, 16 <https://enb.iisd.org/download/pdf/enb12775e.pdf> accessed 21/01/20.

[73] Ibid, 15.

PART V

Concluding matters

10. Conclusions on effective global carbon markets

PERCEIVED PROBLEMS AND CHALLENGES AND THE PROPOSED SOLUTION

Notwithstanding the shortcomings of international emissions trading (IET) under the Kyoto Protocol and its apparent lack of success in reducing overall levels of, or even rates of increases in emissions, a growing number of jurisdictions are implementing mechanisms that put a price on carbon emissions, whether by taxing activities that cause the release of greenhouse gases (GHGs), or by creating markets through which the cost of release of GHGs is internalized to the relevant activities, through emissions trading schemes (ETSs). While the heterogeneity of these mechanisms reflects local preferences, circumstances, capacities, and requirements, to optimize their effectiveness they need to connect into larger, deeper, more liquid markets, which should be less susceptible to manipulation and more effective at generating a stable price for carbon emissions. At the same time, the structure for voluntary cooperative approaches to effect international transfer of mitigation outcomes under Article 6 of the Paris Agreement must be able to account for the global diversity and complexity now evident in national policies and schemes providing for carbon pricing. The corollary is that mitigation outcomes need to be valued, the obvious common metric being their mitigation value, measured in tonnes carbon dioxide equivalent GHG.

This book argues that, in the transition from the Kyoto Protocol to the Paris Agreement, for emissions trading to achieve greater effectiveness it needs to be freed, to an extent, from its current framing in the climate regime. IET – the carbon market as it developed under the Kyoto Protocol – was not designed, but was an outcome of negotiations. It functions as both a climate policy measure and a financial market, operating at an international level. Examination of this duality of purpose indicates imbalances: at the macro level, its genesis resulted in IET, as a financial market, being functionally compartmentalized in the climate policy regime (thus, imbalanced towards the climate function) as a result of which, it is argued, its effectiveness has been impaired; whereas at a granular level, the focus has been on the property rights in what is traded and the entitlements attaching to those rights, rather than in their environmental value (thus, perhaps, being imbalanced the other way). Going forward, it is posited that the carbon market should be framed with a clearer functional identity as a financial market, operating within an equally clearly defined boundary framework of climate change principles and rules. The networking approach proposed to achieve such inter-jurisdictional trading is not only a mechanism for implementing the policy of mitigating GHG emissions, but could be seen also as an opportunity for better integration of international climate change policy, law, and practice into the mainstream political-economy agenda.

The market proposed can be viewed as not a single market, but rather as a trading platform connecting and facilitating transactions between individual, separate markets, each of which continues as an autonomous operation in its own jurisdiction, while participating in the network. The proposal encompasses the digital infrastructure needed to provide the connection between these markets, as well as the legal and administrative structures that will operate, manage, and oversee the network. The proposal consists, therefore, of two distinct elements: first, networking of carbon markets; and second, that networking being carried out using a specific type of digital information technology (IT) architecture, namely, distributed ledger technology (DLT).

Networking is not current practice, presently being only conceptual in nature. The current approach for connecting carbon markets from different jurisdictions is for them to link, which involves alignment of schemes, policies, laws, processes, and so on. This gives rise to political issues, stemming mostly from the perceived impact of system alignment on the sovereignty of the participant jurisdictions. It has been argued that networking better addresses these issues, as the inherent problem of imbalance of negotiating positions (and consequential implications for sovereignty of the disadvantaged party) is far less likely to arise. Networking also holds out a more time efficient process by avoiding the need to homogenize laws, systems, registries, policies, and other elements of the respective participating jurisdictions' systems, thus avoiding lengthy treaty negotiations.

The global recognition of technological developments occurring that fundamentally change how financial services are provided, how markets, business, and government might operate in the future, leads to a conclusion that in proposing a model for networking carbon markets, it is necessary and desirable to address the question of the IT architecture on which the networked market platform could operate. Application of DLT is not without issues, yet at the same time, this technology holds out the promise of useful features, including permanence/immutability of data (supporting traceability, auditability, and robust accounting); decentralized participants (using smart contracting to facilitate transactions directly with reduced need for intermediaries); distributed information sharing and management (enabling balancing of transparency with privacy through the permissioning mechanism); and security (based on hash cryptography, overall design, and the consensus mechanism).

While in some respects, issues addressed by its application might be addressed equally by a well-designed centralized database, DLT does hold out some significant advantages. For instance, while it may be difficult to discern one way or the other in terms of security advantages of different types of cryptography, DL network design, the permissioning and consensus mechanisms, together with the accumulative nature of the ledger entries, suggest a more effective overall security package. Part of that security package is the traceability of traded assets' provenance. This facilitates accounting and audit processes, thereby helping address some requirements of the Paris Agreement. Removing intermediaries from transactions, if achievable – taking into account financial regulations – introduces time and cost savings. Reducing overall end-to-end transaction time, if realizable, would mean that counterparties' risk exposures are reduced, meaning that they would need to make less provision against potential default, consequently releasing more capital for other investment and thereby enhancing the overall efficiency of the market. Other, even more significant aspects are, first, the flexible and relatively simple accessibility that such a network could provide; and second, the facility to access all relevant market and compliance information on the ledger. Realization of these features and advantages is dependent on careful design, both in a technological sense and in terms of the legal and administrative structures that technology supports.

GOVERNANCE STRUCTURE

In proposing this market model, the objective is to arrive at a suitable design for regulatory and institutional frameworks (the governance structure) for trading emissions in the context of the Paris Agreement and wider governing structures, treating emissions trading as a financial market and using DLT architecture to connect different markets in a network. Thus, a framework for analysis for the market has been set out, then applied to analyze the proposed solution, including by mapping it against the current regulatory frameworks.

The governance structure needs to account for the requirements of climate change law; financial markets regulation; and the legal requirements developing in relation to DLT and its applications. Thus, there are three essential elements. Each of these needs to be approached differently as, for example, in relation to climate change there is an existing global governance structure, which allows comparative analysis. On the other hand, financial regulation is principally a domestic law matter, although there is a developing global structure and, accordingly, the proposed market's governance structure has been considered in terms of how it would fit with this developing architecture, while acknowledging the role of domestic regulation. Third, DLT and its applications are recent developments and jurisdictions are currently active in developing their approaches to the challenges posed by the technology and its applications: as such, the approach taken has been to examine the state of these developments and where they might be heading, to assess the compatibility of the proposed governance structure. Finally, the framework for analysis has examined specific legal issues that pertain, in turn, to the technology and its application, to financial market regulation; and to the rules for implementation of the Paris Agreement.

The governance structure dissected vertically discloses three separate, but interacting functional pillars: the first relating to supervisory and regulatory elements; the second relating to the self-regulatory market operation; and the third relating to independently sourced market information, the self-contained mitigation value assessment process. The first pillar features two overriding supervisory bodies, one from the climate policy perspective and the other from the financial market perspective, acting conjointly, ultimately answerable to the Conference of Parties of the UNFCCC acting as the Meeting of Parties to the Paris Agreement (CMA). The self-regulatory market is, in essence, an example of co-regulation, or regulated self-regulation; although the market is operated and regulated by the participating jurisdictions, it would function within a broader regulatory framework, ultimately being answerable to the CMA. The third pillar separates the function of providing market sensitive information, with the aim of ensuring that sources of this information are independent, objective, credible, and reliable, and the process secure and trustworthy. The governance structure viewed horizontally shows seven layers of governance, from the most specific, the rules imposed by jurisdictions on entities they authorize to access inter-jurisdictional trading, up to the broadest, the supervision exercised by the two supervisory bodies, acting conjointly.

Comparing the governance structure for emissions trading under the Kyoto Protocol with that proposed in this book, it is noted that first, the differences in approach under the Paris Agreement to that under the Kyoto Protocol mark a fundamental point of departure. For instance, at the most basic level the terminology under the Paris Agreement suggests a different, possibly less prescriptive approach; second, under the Paris Agreement, there is no developed and developing country party differentiation in terms of specific commitments or obligations and, consequently, the ability to access the trading mechanism; and compliance obligations are replaced by voluntary commitments. A third point of differentiation, reflecting changes in domestic markets such as the European Union, is that the proposed networked market places greater emphasis on the carbon market as a financial market, facilitating better investor protection in areas such as market manipulation or insider dealing, and better investor and market risk management. More significantly, under the Kyoto

Protocol there has been no organized inter-jurisdictional trading platform, as such, whereas the establishment of a trading platform is proposed: thus, the networked market would be largely self-contained and functionally separate from, but capable of supplying required information into, other jurisdictional requirements such as overall emissions accounting and reporting. Finally, the networked market proposed is based on continuing, autonomous operation of the carbon markets in the participating jurisdictions, thus compliance and enforcement will primarily remain a domestic jurisdictional matter thereby, it is anticipated, being both more likely and more effective. It is concluded that, notwithstanding the networked market being largely self-contained and functionally separate from the jurisdictions participating, it would be only an adjunct, or supplement, to their domestic mitigation activities – a tool to which they may have recourse, as they see necessary, in fulfilling their nationally determined contributions.

A key aspect of the approach for global financial market governance is that, as a supplement to sound micro-prudential and market integrity regulation, national financial regulatory frameworks be reinforced with a macro-prudential overlay to promote a system-wide approach to financial regulation and oversight and to mitigate the build-up of excess risks in the system. Thus, supervision over commercial actors in financial markets should be based on a two-tier system with national supervisors continuing to exercise micro-prudential oversight and a level of macro-prudential oversight introduced for financial markets as a whole, in order to provide early recognition of systemic risks. The governance structure proposed for the networked carbon market reflects a similar two-level approach.

In terms of the technology, the historical evolution of the use cases and application of DLT from an alternative payment system (bitcoin) to the present emphasis on initial coin offerings and investment, points to a changing and increasing array of risks for both users and regulators to countenance, while remaining focused on the potential and opportunities technological innovation can provide. As a consequence, regulatory approaches and responses cover the gamut from wait-and-see, through supportive measures and guidance vis-à-vis current laws, to sandboxing; innovator-friendly regulatory frameworks, technology neutral risk-based application of anti-money laundering/counter terrorism financing (AML/CTF), consumer protection, and systemic risk management provisions. Surveys of jurisdictions indicate there are very divergent approaches to crypto-assets, making it difficult to discern any stand out direction that regulation might take apart from, perhaps, increasing in amount. All the same, the obvious starting point from a regulatory perspective has been for jurisdictions to consider whether the financial sector applications of DLT fall within the ambit of existing financial regulation. The analysis devolves into a question of the nature of the token issued, its economic purpose and function and the rights and entitlements, if any, attaching to it.

The question of the nature of tokens, issued on distributed ledger platforms used in financial applications, is at the intersection between legal issues relating to applications of DLT and financial market regulation. Further, if the units traded on the network of markets – that is, as tokens traded on the DLT platform – represent units of mitigation value (MV), then the regulation of applications of DLT and financial market regulation intersect also with climate law considerations. The nature of the 'mitigation outcome/financial instrument/token' is germane to all three areas of law considered and thus lies at the centre of this analysis. Consideration of the mechanism by which transactions might take place inter-jurisdictionally on the networked market leads to the question of the need for a transaction unit, as a vehicle for effecting transactions. The assumption that a transaction unit is more than a notional step in the transaction process, and capable of having a separate legal identity as a store of value, leads to the conclusion that it would be of the same legal nature as the mitigation outcome from which it derives, but with a standardized value. As the legal nature of mitigation outcomes depends on their domestic source jurisdiction, an analysis tree for determining domestic regulatory treatment of a transaction unit is proposed. All the same, there would need also to be a way of determining the law applying to the transaction units on the network platform (that is, externally to any particular domestic jurisdiction), for which various approaches are suggested, such as applying the governing law of the network platform, or as might be provided in the network platform's constitutional document.

While continuing research into potential design requirements for the application of distributed ledger technology to a network of carbon markets is needed and is ongoing, it is concluded that a market designed along the lines proposed as a mechanism for implementing GHG mitigation policy allows pursuit of climate policy objectives, giving effect to the elements of Article 6, Paris Agreement. It allows proper and efficient operation of the market by introducing appropriate elements of financial regulation and provides for and is responsive to technological developments. Thus, a design is arrived at for regulatory and institutional frameworks for trading emissions considered suitable in the context of the Paris Agreement and wider governing structures.

The logic for this governance structure is the separation and clarity of functions; flexibility; legal certainty, with independence and objectivity; and the self-reinforcing balancing of the dual functions of the carbon market – effective market operation promoting the climate objective and driving higher mitigation ambition, with market self-regulation (co-regulation) promoting greater efficiency and more effective operation. The main elements are:

- the functional separation into three pillars being regulatory/supervisory functions; self-regulatory (co-regulatory) market operation; and the independent provision of market information;
- second, the flexibility of the networking structure on a distributed ledger platform fostering a level-playing field and that jurisdictions can join or leave, efficiently and relatively easily, according to their assessment of the domestic economic needs and perceived benefits; and
- finally, the use of the DLT architecture for the trading platform to promote robust accounting and as a way of ensuring environmental integrity.

It has long been acknowledged that private sector engagement is essential for success of the UNFCCC processes, including constructive participation in carbon market mechanisms, as a way of driving investment in low-carbon technologies.[1] The scientific evidence is that limiting global warming to 1.5°C as opposed to 2°C will lower impacts, but necessitates more immediate and greater efforts to mitigate emissions.[2] This is especially so given, on one hand, the investment timeframes for relevant critical infrastructure and, on the other, the time lags before climate impacts ameliorate. Engagement of the private sector at scale is essential if the process for cooperative approaches involving the international transfers of mitigation outcomes, under Article 6 Paris Agreement, is to achieve enhanced mitigation. As the IPCC notes with ever increasing frequency, time for action is running short if dangerous anthropogenic climate change is to be avoided. The approach outlined here holds out the potential of a shorter route to implementation. At the same time, while the application of DLT might disintermediate some financial transaction counterparties, it should not, of itself, alienate relevant climate stakeholders, such as the CMA, Secretariat/ITL, or subsidiary bodies, whose roles would be reinforced.

---

[1]   For example: WBCSD Secretariat, Ecofys and Climate Focus, 'Private Sector and the UNFCCC Options for Institutional Engagement', Final Report 31/8/2010 <https://www.wbcsd.org/Clusters/Climate-Energy/Resources/Options-for-institutional -engagement-in-the-UNFCCC-process> accessed 11/04/18; European Bank for Reconstruction and Development (EBRD), Operationalising Article 6 of the Paris Agreement: Perspectives of developers and investors on scaling-up private sector investment, May 2017 <www.ebrd.com> accessed 21/09/17.

[2]   Intergovernmental Panel on Climate Change (IPCC), *Global Warming of 1.5°C*, Special Report, Summary for Policymakers, 2018 <http://report.ipcc.ch/sr15/pdf/sr15 _spm_final.pdf> accessed 09/11/18.

MATTERS BEYOND THE SCOPE OF THIS BOOK

The reader of this work will appreciate the large breadth of subject matter, disciplines, laws, legal concepts, and scope that needs to be spanned in order to deal with a topic such as climate change and, in particular, the construction of mechanisms to implement policies that address it. Nevertheless, some matters are beyond the scope of this book. The first is the mitigation value assessment process, only briefly dealt with[3] as one of the operational mechanisms that will be required for the proposed networked market to operate. All the same, the importance of this process is reflected in the governance structure in the third pillar, the independent source of market information, described in Chapter 8.

Additionally, there are several matters of a technical nature, which while noting that they exist and may require attention in the course of further development of the proposals set out in this book, do not need to be explored or resolved conclusively here. First, there is no need, nor is it possible at this point in time, to reach a definitive conclusion as to the technological advantages of a distributed IT architecture over a centralized approach. This depends to a significant degree on the particular use case under consideration and to design elements. Second, issues of cybersecurity and the potential for human error are ever present. While the application of DLT to a networked carbon market includes a description of the perceived security benefits, it is recognized that this provides only a high-level coverage of what is both an intricately detailed and highly technical subject. Finally, the consideration of legal issues associated with the distributed ledger technology mentions liability for technical default issues, such as incorrect coding or systems failures. These have not been explored in any detail, as technical defaults can affect any system or technology and so are not particular to distributed ledger technology.

---

[3]   Chapter 7.

Appendix[1]

Example rules:

(i) Rules governing market operation and jurisdictional participation

In order for the distributed network to create the framework for an inter-jurisdictional emissions trading market that conforms to international (and national) climate change policy, certain fundamental rules and principles are of critical importance. These could include:

(a) a condition that, in all transactions, environmental integrity is protected and preserved, that is, not compromised or reduced;[2] in other words, participation in the distributed network market must always contribute to an overall reduction, or at least no increase, in greenhouse gas emissions for jurisdictions with which the counterparties to a transaction are connected, and for the networked market overall;

(b) application of the supplementarity principle,[3] dictating that not more than, say, 25 per cent of the units held on a domestic ETS registry may be sourced from the international market (that is, from sources outside the particular jurisdiction), although individual jurisdictions might be free to set limits even lower than this (which would then be enforced through smart contract code terms applicable to entities from that jurisdiction);

(c) the requirement that any individual trading actor (in effect, a trading entity within a given participant jurisdiction's ETS) retain a compli-

---

[1] Previously published in: Justin D. Macinante, 'A Conceptual Model for Networking of Carbon Markets on Distributed Ledger Technology Architecture' [2017] *CCLR* 243, 252–3. Reproduced with permission.
[2] Such a rule is clearly necessary, and its enforcement a high priority, given the terms of the Paris Agreement.
[3] The supplementarity principle as stated for the purposes of the Kyoto Protocol is that '… the use of the mechanisms [International Emissions Trading, CDM, JI] shall be supplemental to domestic action and that domestic action shall thus constitute a significant effort made by each Party included in Annex I to meet its quantified emission limitation and reduction commitments under Article 3, Paragraph 1'. (Article 1 Draft Decision -/CMP.1 (Mechanisms) contained in Decision 15/CP.7, Marrakech Accords). The 25 per cent figure here is only for the purposes of indicating application of the principle in this context.

ance reserve of, say, 75 per cent of its emissions-related obligations;[4] it is noted that Parties in Annex B under the Kyoto Protocol had compliance reserve commitments under the rules for international emissions trading; (following a principle of economic/compliance risk management comparable with, for instance, the minimum capital deposits of banking institutions);

(d) application of an automatic and immediate block on all transactions involving entities from a jurisdiction, where that jurisdiction indicates an intention to withdraw from the distributed network;

(e) application by a jurisdiction of national rules of acceptance/exclusion prescribing 'mitigation value' (MV) limits in regard to other jurisdictions with which it would be willing to permit entities authorized by it to trade; for example, a refusal to permit transactions with a jurisdiction whose MV is below (or above) a specified level, or (whose MV) is outside a specified range;

(f) upon a jurisdiction joining the distributed network, the provision and maintenance of a surety in respect of its contribution to joint network costs, to be forfeited, for instance, if the jurisdiction were to withdraw without proper notice; transaction rules would correspondingly block transactions by entities from any jurisdiction not providing an adequate surety;

(g) the possibility of jurisdictional adjustment of trading maxima or minima under rules (b) and (c) (by respectively lowering or raising the figure) as they apply to a given jurisdiction's own compliance entities (i.e., traders within its own domestic ETS), for the purpose of managing domestic market activity; and

for those rules allowing for jurisdictions to make adjustments vis-à-vis trading activity of entities authorized by them (for example, rules (b) and (c) above), an appropriate notification procedure would need to apply.

(ii) Rules governing transactions

Trading within the market established through the distributed network could be possible only where the individual actors (traders) conform to a minimum set of standardized rules or principles. These might include, at least, the following:

(a) the seller must, in fact, hold the trading units offered for sale, evidenced by a registry/ledger entry;[5] this requirement might be satisfied

---

[4] Hence, this would apply only to those trading entities that have compliance obligations in the jurisdiction, so not to say, brokers or market makers.

[5] In other words, there would be no scope for short-selling.

by the seller's ability to convert the units offered into transaction units (TUs);

(b) the buyer must hold the funds necessary to complete the transaction, evidenced by bank records, automated bank confirmation, or deposit of the requisite amount into an account accessible to the seller on settlement;[6]

(c) automatic application of the conversion/discount rate, where relevant, between the jurisdictions concerned, applicable at the time of the transaction or, in the case of TUs, conversion by the seller of its trading units into TUs (at the applicable rate), and, upon the price money being available/transferred, either the transfer of the TUs, or the conversion of the TUs into the buyer's domestic units (at the applicable rate) and the transfer of these to the buyer's account in its domestic ETS registry; and

(d) on settlement, as per (c), updating of all copies of the ledger.

The 'smart contract' would operate on the basis that, if any term or condition essential for such a transaction were not met, the transaction would not proceed. Thus, unless all such requirements are satisfied within some predetermined period (say, within a specific number of hours of the initiation of the transaction), the transaction would fail and lapse.

Essential terms and conditions might typically include the following information and specifications:

- name and jurisdiction of seller;
- domestic authorization, satisfactory KYC[7] and AML[8] checks on seller;
- name and jurisdiction of buyer;
- domestic authorization, satisfactory KYC and AML checks on buyer;
- certification or proof that the transaction is accepted as not negatively impacting upon environmental integrity;[9]
- certification that the transaction would not cause the buyer's jurisdiction to breach the supplementarity principle (noting that either jurisdiction may have the level set lower than maximum applicable in the distributed system as a whole);

---

[6] An alternative would be that the smart contract would not permit the transfer of the transaction units from seller to buyer until the money was either available for or, in fact, had been transferred to the seller's bank account.

[7] Know-your-customer.

[8] Anti-money laundering.

[9] As noted earlier, most likely this would need to be part of the MV setting process.

- certification that the transaction would not cause either the seller or the seller's jurisdiction to breach the compliance reserve (noting that the level may have been set higher than the minimum required for the distributed system as a whole);
- that the conversion rate, where relevant, is acceptable to the buyer's jurisdiction;
- that both jurisdictions have provided and maintain an acceptable surety in regard to their financial obligations towards the operation of the distributed network;
- confirmation that the seller holds and is entitled to sell the domestic units offered for sale;
- confirmation that the buyer has funds to complete transaction; and
- where relevant, the application of the correct conversion rate between jurisdictions or, for TUs, between each jurisdiction and a TU.

Once all such terms and conditions are satisfied, the transaction would proceed automatically and irreversibly.

Index